# SpringerBriefs in Geoethics

**Editor-in-Chief**

Silvia Peppoloni, National Institute of Geophysics and Volcanology (INVG), Rome, Italy

**Series Editors**

Nic Bilham, University of Exeter, Penryn, UK

Peter T. Bobrowsky, Geological Survey of Canada, Sidney, Canada

Vincent S. Cronin, Baylor University, Waco, USA

Giuseppe Di Capua, National Institute of Geophysics and Volcanology (INVG), Rome, Italy

Iain Stewart, University of Plymouth, Plymouth, UK

Artur Sá, University of Trás-os-Montes and Alto Douro, Vila Real, Portugal

Rika Preiser, Stellenbosch University, Stellenbosch, South Africa

SpringerBriefs in Geoethics envisions a series of short publications that aim to discuss ethical, social, and cultural implications of geosciences knowledge, education, research, practice and communication. The series SpringerBriefs in Geoethics is sponsored by the IAPG—International Association for Promoting Geoethics (http://www.geoethics.org).

The intention is to present concise summaries of cutting-edge theoretical aspects, research, practical applications, as well as case-studies across a wide spectrum.

SpringerBriefs in Geoethics are seen as complementing monographs and journal articles, or developing innovative perspectives with compact volumes of 50 to 125 pages, covering a wide range of contents comprising philosophy of geosciences and history of geosciences thinking; research integrity and professionalism in geosciences; working climate issues and related aspects; geoethics in georisks and disaster risk reduction; responsible georesources management; ethical and social aspects in geoeducation and geosciences communication; geoethics applied to different geoscience fields including economic geology, paleontology, forensic geology and medical geology; ethical and societal relevance of geoheritage and geodiversity; sociological aspects in geosciences and geosciences-society-policy interface; geosciences for sustainable and responsible development; geoethical implications in global and local changes of socio-ecological systems; ethics in geoengineering; ethical issues in climate change and ocean science studies; ethical implications in geosciences data life cycle and big data; ethical and social matters in the international geoscience cooperation.

Typical topics might include:

– Presentations of core concepts
– Timely reports on state-of-the art
– Bridges between new research results and contextual literature reviews
– Innovative and original perspectives
– Snapshots of hot or emerging topics
– In-depth case studies or examples

All projects will be submitted to the series-editor for consideration and editorial review.

Each volume is copyrighted in the name of the authors. The authors and IAPG retain the right to post a pre-publication version on their respective websites.

The Series in Geoethics is initiated and supervised by Silvia Peppoloni and an editorial board formed by Nic Bilham, Peter T. Bobrowsky, Vincent S. Cronin, Giuseppe Di Capua, Rika Preiser, Artur Agostinho de Abreu e Sá, Iain Stewart.

Giuseppe Di Capua · Luiz Oosterbeek
Editors

# Bridges to Global Ethics

Geoethics at the Confluence of Humanities and Sciences

 Springer

*Editors*
Giuseppe Di Capua
International Association for Promoting
Geoethics (IAPG)
Istituto Nazionale di Geofisica e
Vulcanologia
Rome, Italy

Luiz Oosterbeek
International Council for Philosophy
and Human Sciences (CIPSH)
Geosciences Centre
Instituto Politécnico de Tomar
Tomar, Portugal

ISSN 2662-6780        ISSN 2662-6799   (electronic)
SpringerBriefs in Geoethics
ISBN 978-3-031-22222-1      ISBN 978-3-031-22223-8   (eBook)
https://doi.org/10.1007/978-3-031-22223-8

This Springer imprint is published by the registered company Springer Nature Switzerland AG
The registered company address is: Gewerbestrasse 11, 6330 Cham, Switzerland

# Acknowledgement

We wish to thank Silvia Peppoloni for having suggested the general idea of this editorial project and proposing to develop it within SpringerBriefs in Geoethics.

# Contents

# Chapter 1
# Introduction to the Book "Bridges to Global Ethics"

Giuseppe Di Capua and Luiz Oosterbeek

**Abstract** This book wants to enrich the current discussion on geoethics and global ethics within the geoscience and humanities communities, providing new contents and insights elaborated by scholars with different disciplinary backgrounds.

**Keywords** Geoethics · Global ethics · Geosciences · Humanities · Earth system

## 1.1 Motivation of the Book

Anthropogenic global changes of social-ecological systems and their environmental, social, and cultural implications need to be discussed and faced by common grounds in order to make choices and take decisions capable to give planetary answers to problems that do not know economic, religious, gender, or national boundaries.

In this perspective, decision-making and problem-solving should balance the expectations of finding common decisions for the benefit of humanity and the necessity to enhance diversity of approaches, methods, and strategies belonging to different socio-cultural communities. An inclusive process should be favoured and oriented by planetary horizons to be defined through a convergence of understandings rooted in people's practices and perceptions, built on different cultural and institutional backgrounds and diverse systems of belief, and recognising that a diversity of goals is compatible with a convergence of those understandings, fostering a global and peaceful transformation.

G. Di Capua (✉)
Istituto Nazionale di Geofisica e Vulcanologia, Rome, Italy
e-mail: giuseppe.dicapua@ingv.it

International Association for Promoting Geoethics—IAPG, Rome, Italy

L. Oosterbeek
Geosciences Centre, Instituto Politécnico de Tomar, Tomar, Portugal
e-mail: loost@ipt.pt

International Council for Philosophy and Human Sciences—CIPSH, Tomar, Portugal

Scientific, technological, and scientific progresses have undoubtedly produced numerous advancements for humanity, but, as already highlighted by the philosopher Hans Jonas (1903–1993), those have occurred without an objective ethical framework to serve as a guide. The geo-ecological alterations in the last two centuries and especially the "great acceleration" in the deterioration of the Earth system and the human habitat basic requirements raise the need for convergent human actions, which implies to consider the dimensions of meaning and assignment of value in relation to problem-solving. This may lead to bridge the ethical gap in the human development that is blocking adequate human adaptive sustainable responses to environmental changes, moving apart from a merely quantitative approach to the development of human societies that has not considered cultural, social, and environmental disruptive implications of those changes. One challenge that emerges from this historical and global context is the need of improving cross-cultural understandings and building convergent aims, and one important way of doing so is to pursue the effort of building a global ethics rooted in different cultural traditions. Within such path, understanding the Earth as a web of meanings and interactions at different layers is fundamental, namely through the identification of human sustainable practices and reflections, since these are the expressions of ethical understandings grounded in Human–Environment and Human–Human relations.

## 1.2  Aim of the Book

Some recent publications sum up the development of geoethics in the last ten years, provide a theoretical framework and a scheme, and propose geoethics as a global ethics to face grand challenges for humanity (Di Capua et al., 2021; Peppoloni & Di Capua, 2020, 2021, 2022).

Geoethics has been defined as the *"research and reflection on the values which underpin appropriate behaviours and practices, wherever human activities interact with the Earth system"* and puts at the centre of its reflections and practical applications the following ethical principles:

- the dignity of all entities constituting the planet to give new roots to conservation policies to save the habitability of the Earth system for human species while respecting the intrinsic value of any other living and non-living entity;
- the responsibility and freedom as criteria of acting within the four domains of human experience (individual, interpersonal/professional, societal, environmental).

Selected authors (Silvia Peppoloni, Giuseppe Di Capua, Tanella Suzanne Boni, Robert Frodeman, Sofia Belardinelli, Telmo Pievani, Harold Sjursen, Luiz Oosterbeek, Rosi Braidotti, Torbjörn Loden) provide specific disciplinary interfaces of ethics (Earth and life sciences, humanities, social sciences, including the relation between materiality and perceptions and between the anthropocentric programme of modernity and current societies) and a final chapter that resumes the reflection of global ethics, contributing to further developments of the geoethical thought.

# References

Di Capua, G., Bobrowsky, P. T., Kieffer, S. W., & Palinkas, C. (2021): Introduction: geoethics goes beyond the geoscience profession. In G. Di Capua, P. T. Bobrowsky, S. W. Kieffer, C. Palinkas (Eds.), *Geoethics: status and future perspectives*, SP508, Geological Society, (pp. 1–11). https://doi.org/10.1144/SP508-2020-191

Peppoloni, S., & Di Capua G. (2020). Geoethics as global ethics to face grand challenges for humanity. In G. Di Capua, P.T. Bobrowsky, S.W. Kieffer, C. Palinkas (Eds.), *Geoethics: status and future perspectives* (Vol. 508, pp. 13–29). Geological Society of London, Special Publications. https://doi.org/10.1144/SP508-2020-146

Peppoloni, S., & Di Capua, G. (2021). Geoethics to start up a pedagogical and political path towards future sustainable societies. *Sustainability, 13*(18), 10024. https://doi.org/10.3390/su131810024

Peppoloni, S., & Di Capua, G. (2022). *Geoethics: Manifesto for an ethics of responsibility towards the Earth* (pp. XII+123). Springer, Cham, ISBN 978-3030980436. https://doi.org/10.1007/978-3-030-98044-3

# Chapter 2
# Geoethics for Redefining Human-Earth System Nexus

Silvia Peppoloni and Giuseppe Di Capua

**Abstract** The globalized society is held together by an intricate system of human relationships. This system constitutes a planetary architecture characterized by (a) a complex technological structure, (b) the homogenization of cultural forms and economic systems, (c) growing social, political and economic inequalities. Faced with planetary systemic perturbations (such as pandemics and wars), the globalized society shows criticalities, but also strengths, despite it is still too vulnerable to anthropogenic environmental changes. These changes modify the physical–chemical-biological characteristics of the Earth system and therefore represent a great threat to human communities, more serious than the pandemic threat from SARS-CoV-2, perhaps equal to the threat of a nuclear war, since it is the habitability of the planet by humanity and many other living species to be in danger. In order to promptly address the dangers of the anthropogenic changes underway, a closer and more structured international cooperation between states is needed. There are no alternatives. But this goal today appears increasingly difficult and distant due to the international geopolitical instability triggered by the war in Ukraine. In fact, only, human communities that share ethical principles and values on which to base new forms of relationship between human beings and the Earth system are able to face the planetary ecological crisis and build a possible future on Earth. In this perspective, geoethics is proposed as global ethics of a complex world, founded on the principles of dignity, freedom and responsibility and aimed at the renewal of the human-Earth System nexus and the realization of an ecological humanism.

**Keywords** Geoethics · Earth system · Anthropogenic global changes · Ecological humanism · Global ethics

S. Peppoloni (✉) · G. Di Capua
Istituto Nazionale di Geofisica e Vulcanologia, Rome, Italy
e-mail: silvia.peppoloni@ingv.it

G. Di Capua
e-mail: giuseppe.dicapua@ingv.it

International Association for Promoting Geoethics—IAPG, Rome, Italy

## 2.1 Introduction: Globalized Society, Anthropogenic Changes and Biocidal Loops

The globalized society is a real object made up of infrastructural networks, through which mass and energy flows move on the surface and in the atmosphere of the planet to support human communities. Infrastructural networks are articulated into systems of commercial, digital, tourism, military, industrial, cultural, social relations, which allow rapid transfers of tangible and intangible assets consisting of technologies, goods, ideas, symbols, languages, world views, as well as fears and hopes. In our time, this phenomenon of interdependence has assumed a dimension and a pervasiveness without equal in history and has produced a globalization of needs, expectations and demands through a process that nevertheless Edgar Morin (2020) defines as *"without solidarity."*

Both due to the close connection between the parties and the speed of transfer within the structured networks, the SARS-CoV-2 pandemic, as well as the war in Ukraine, has highlighted how vulnerable the globalized society is to a perturbation. Indeed, within these dense planetary networks, a locally generated socio-natural disturbance (a war, a volcanic eruption, a pandemic, an accident to a technological infrastructure, a coup, a financial crisis) can evolve in ways not easily predictable and have effects that can rapidly propagate in human ecosystems through relational networks, up to having repercussions on a planetary scale in a very short time.

However, the contingency of the SARS-CoV-2 pandemic has also highlighted a positive aspect of the globalized society, which has shown a good capability to react to the "disturbance" caused by the spread of the virus. This reaction has resulted in the rapid development of vaccines, thanks to the immediate circulation of scientific information and the international cooperation of research groups, the technological and industrial capabilities of the most advanced countries in the production and distribution of vaccine vials, the timely government action in managing a vaccination campaign on a scale never experienced before. On the other hand, we cannot fail to mention the "vaccine nationalism" (Bollyky & Bown, 2020) and the inequity in the distribution of vaccines, unequivocal phenomena of the persistent inequalities that characterize global human society and the relations of power and domination over weakest ones existing within and among countries.

Human communities, supply (or value) chains, connection networks are just some of the elements that make up the human ecological niche that is the space that human beings are continuously shaping incessantly from thousands of years (Ellis et al., 2021), together or in contrast with nature, in order to inhabit the Earth system, modifying, often unconsciously, ecosystems and social-ecological dynamics according to complex mechanisms, in a continuous action-reaction dyad. Climate change is only one of the ongoing colossal anthropogenic processes of modification of the human niche (Meneganzin et al., 2020), which affect humanity by directly and indirectly modifying its perspectives, expectations and decisions. Wuebbles (2020) states that *"The rapid changes occurring in the Earth's climate are impacting every part of our lives, economically, culturally and physically, and these are driving concerns about vulnerability, equity and justice.*" If for some decades, the study about global warming has concerned only a part of the scientific community, committed to quantifying its entity and evaluating its possible consequences on the physical–chemical systems of the planet, more recently the social, economic, ethical and cultural implications that such phenomenon brings with it have become the object of study of various other disciplinary fields. This is demonstrating on the one hand the complexity of the natural phenomena taking place; on the other hand, the growing awareness by experts from different fields of knowledge on the fact that the anthropogenic modifications of the natural environment are not limited to the extensive impact on geo-ecosystems (ensembles of biotic and abiotic systems), but they also affect the systems of human relations, modifying their evolutionary trajectories, which in turn trigger further environmental changes. This process of continuous adaptation to persistent imbalance is a natural fact, but *Homo sapiens* has accelerated it to the point that human-induced changes in the Earth system seem to occur much more rapidly than human changes necessary to adapt to changing external environmental conditions. Humanity may therefore no longer be able to adapt to the environmental changes it has itself produced, entering a dangerous biocidal loop that may only be interrupted by sudden and providential scientific discoveries and technological applications.

It is in this scenario that concerns must be placed on the possible use in the near future of climate geoengineering (which aims to counteract global warming by subtracting $CO_2$ from the atmosphere or increasing albedo) and deep sea/ocean mining (aimed at obtaining large quantities of minerals necessary for the transition to an economy that does not use fossil fuels). These technologies pose no longer negligible ethical questions and dilemmas, which must be addressed on a scientific and philosophical basis, with interdisciplinary and transdisciplinary approaches and without ideological prejudices, since they do not concern only technical or ecological aspects, but the future life of the human species on Earth (Peppoloni & Di Capua, 2020).

## 2.2 Pandemic and Global Warming as Effects of the Human Crisis

The basic element that links the pandemic and global warming is the unhealthy relationship that humanity has established with the natural environment, albeit with great differences between the various cultures. If we refer to Serres (2019), humanity has transformed from "prey" to "predator". However, in the near future, the great human-induced planetary geophysical upheavals risk making humanity succumb, making it go back to being "prey" in nature.

In Western culture, the dichotomy existing between human beings and nature has distant roots (Capra, 1975, pp. 20–21). For some, it may highlight a form of deviance of the human species from an alleged ecological harmony, which is reflected in the myth of "Wilderness" (Denevan, 1992, 2011; Ellis et al., 2021; Fletcher et al., 2021).

However, it is reasonable to think that the dominance of selfish attitudes of the individual and of human groups, which have their own evolutionary justification, over time has led human beings to a vision of the other from themselves as an object to be dominated and not as the subject of a relationship to be built to guarantee one's existence within a systemic equilibrium. The use of technology by humans has gradually become an exercise of omnipotence, the exaltation of human domination over a nature perceived as a constant threat, to be kept at bay to control its dynamics.

The split between humans and nature seems to have widened to the point of determining a distancing of the human being from their own humanity, up to the extreme case of the creation of a virtual life within the metaverse that set of virtual spaces in which the individual is integrated and personified in an avatar, and as such, he/she moves, establishes relationships, acts, abandoning the physical–biological dimension to live in a digital dimension.

In this process, the dematerialization of the bond with physical places takes place (Ferrarotti, 2009), and the crushing of the times of human experience on the present. The causality links between the phenomena of nature dissolve, the dynamics of which often take place over a long period of time, far beyond the common human perception (see Frodeman's, chapter in this volume, 2023). In this indefinite and liquid reality, disconnected from space and time, the human being finds themselves living an abstraction, without connection with material, tangible nature, without sensory perception of phenomena and contact with places, ecosystems, geodiversity and biodiversity. It can be said with Serres (2019) that *"we have lost the world,"* even if sudden events such as a war can bring us back to the materiality of our existence.

Technological development has been both the effect and the cause of this change in the life of humanity. Technology continues to oscillate constantly between two polarities: on the one hand, it is the extraordinary tool, produced by human creativity at the service of the material and cultural progress of our societies; on the other hand, it is a formidable picklock capable of undermining the authentic humanity of the human being and favouring their engineered homogenization (cyborgization), implementing that will to power and prevarication of the human over the human (an example is the atomic bomb) and of the human on nature (as occurs in the alteration of biogeochemical cycles).

Ever since the end of the seventies of the last century, for the German philosopher Hans Jonas (1903–1993) was clear that technology without ethics was a risk for humanity. Jonas calls upon society to a new commitment: that of worrying about the consequences of technological developments for future generations. The principle of responsibility (Jonas, 1979) is the proposal to anchor human action to an ethical criterion in a technologized world, thus releasing it from an operational logic forced on the present. For Jonas, modern human capacities to manipulate the planet, with destructive effects even in the long term, require an action whose consequences allow the continuity of human life, and therefore of future generations, on Earth.

In this perspective, the governments and public opinions that have followed one another over the years (we refer mainly to the Western ones in a broad sense, which generated the ecological crisis through the predominant socio-economic models) have not had the same ability as Jonas to look to the present and even less to the future of a rapidly changing world. The dizzying technological development of the last few decades has been used to fuel a new economic phase in perfect continuity with the past (digitization also has its important environmental costs, as well as part of the so-called "green economy"), without also being accompanied by a profound cultural, ethical, social and environmental reflection on its own consequences and on an idea of human progress that went beyond limited interests and reductionist analyses.

It is evident that the ontological fracture between human being and nature must find a solution. Serres (2019) had proposed to sign a natural contract that would put an end to the atavistic war between the human world and the natural world by regulating the forms of conflict.

The anthropogenic changes cannot be faced only with science and technology, just as their management cannot be delegated exclusively to those small political, economic and financial elites that often caused them. These phenomena have a profound impact on the entire globalized society and therefore require responses from the global human community, through the sharing of knowledge, experiences, proposals, ideas and techniques.

Anthropogenic global warming with its many associated phenomena is a scientifically ascertained fact (Head et al., 2021; IPCC, 2021; Jouffray et al., 2020; Ripple et al., 2020, 2021), and some governments have been aware of it since the 1960s. Although the scientific community had already reported its seriousness and attributed the causes to carbon dioxide ($CO_2$) emissions deriving from the use of fossil energy sources (The White House, 1965), however, until a few years ago, the topic of global warming had been de facto excluded from political agendas. Only, in the last ten years has it made a strong comeback in the public debate thanks to the commitment of the environmental and youth movements. On the contrary, the environmental issue has been heavily considered in commercial marketing to organize greenwashing campaigns, which focused on the concept of sustainability to convey new sales messages.

Today, planetary environmental issues have finally been included among the political issues to be debated at the highest level of relations between governments, albeit severely slowed down by other global issues such as the pandemic and military conflicts. But, the beginning of a real modification of the economic, financial, social, cultural and political paradigms, responsible for the excessive anthropogenic pressure on the environment and the deterioration of ecosystems, still seems far away.

The actions implemented by the States to change the development trajectory (such as the Paris Agreement of COP21[1] or the decisions taken during COP26[2]) are insufficient, mostly dictated by wait-and-see logics that seem to hope for future saving interventions of scientific research and technology, rather than starting the construction of a more sustainable and respectful future for social-ecological systems with a sense of responsibility.

---

[1] https://unfccc.int/process-and-meetings/the-paris-agreement/the-paris-agreement (accessed 13 July 2022).

[2] https://ukcop26.org/the-conference/cop26-outcomes/ (accessed 13 July 2022).

Unfortunately, the current ecological crisis is serious, and the parameters that scientists use to track its trend over time do not give rise to ambiguity (Callaghan et al., 2021; Head et al., 2021; Jouffray et al., 2020; Ripple et al., 2020, 2021). And on closer inspection, it is primarily a crisis of the human being, a crisis of civilization, especially of the Western one, determined by two factors (Peppoloni & Di Capua, 2021a):

(a) the irresponsibility of the ruling classes in pursuing development goals without taking into account the environment and social inequalities;
(b) the lack of responsibility of individuals towards common issues, fuelled by a culture of success that has promoted selfishness without balancing it with solidarity.

These two factors are linked and also highlight an apparent paradox. Indeed, it was the power elites who chose to direct society on trajectories of ecosystemic unsustainability, inducing the modification of individual behaviours and amalgamating needs and expectations in the civilization of mass consumption (Bonneuil & Fressoz, 2019). On the other hand, once the environmental problem has exploded in all its drama, it was the large multinational corporations of energy and commerce, primarily the oil ones that began to promote initiatives aimed at diverting attention from their responsibilities towards global warming, to focus on the irresponsibility of individual consumerist behaviours (Oreskes & Conway, 2019).

In this complex framework, ethics must be rethought as an indispensable tool on which to base the regeneration of human society, to make it more inclusive, equitable, peaceful, socially supportive and ecologically oriented in its choices.

In our vision, this regeneration project must be based on the individual, as advocated by Morin (2020). The individual is the quantum of social construction, the fundamental particle that can become a conscious and not forced source of that spirit of solidarity that must unite humanity in the search for a more responsible, just and respectful way of life towards human beings and the environment. It is on the individual that action must be taken to enhance those human qualities that make it possible to build supportive relationships and a peaceful co-existence.

But, to do this, it is necessary to take into account the following:

(1) The ecological crisis is a planetary issue that will develop over several generations.
(2) The set of human beings and the socio-cultural, economic, engineering and technological structures that they produce exerts a strong pressure on ecosystems and natural resources.
(3) Geoscientific knowledge is an indispensable tool for managing the functioning of modern globalized societies, poorly considered in decision-making processes. Human communities need adequate geoscientific literacy since geosciences can help redefine the relationships between humans and nature on the basis of some scientific discoveries that outline their philosophical value for human life by modifying the perception that humans have of their interaction with other non-human entities: deep time, the systemic dimension of planetary relationships characterized by feedback loops processes, the grandeur and

power of geophysical phenomena (Cervato & Frodeman, 2012; Frodeman, 1995, 2014; Stewart, 2022; Zen, 2018).

(4)   Humanity does not take into account planetary boundaries and systemic sustainability thresholds (Rockström et al., 2009; Steffen et al., 2015) in building their own living space (the anthroposphere).

(5)   The anthroposphere has now come to coincide with the physical limits of the planet. Consequently, the human ethical space has also expanded to a planetary dimension.

(6)   Over the next few decades, the persistent overcoming of ecological tipping points will induce substantial changes to the Earth system, which could prove incompatible with maintaining the planet's habitability for the human species (Lenton et al., 2019; Steffen et al., 2018).

(7)   The ecological crisis is also the effect of the enormous political, economic and social imbalances existing on Earth: facing it in a conscious and effective way implies having to reconsider at the same time political structures, economic models of reference, social issues and relationships between States. There are inequalities in human access to resources, both nationally and internationally, which fuel the ecological crisis and that the ecological crisis in turn affects.

(8)   Solutions to global environmental problems can only be sought through widespread international cooperation. Global anthropogenic changes do not know boundaries and therefore require concerted responses on a planetary level. For this reason, in the globalized world, there can be no solution to a problem without sharing and solidarity.

(9)   The ecological crisis and the social problems that derive from it or that cause it require scientific knowledge and the sharing of multidisciplinary knowledge. There is no effective action to find answers to social-ecological problems that does not consider multidisciplinary and transdisciplinary approaches.

(10)  The search for answers and common solutions presupposes the development of a planetary consciousness shared by the various human communities, through the creation of a reference framework of ethical principles and values on which to base the sense of belonging to a "community of destiny "(Morin, 2015, 2020). By community of destiny, we mean that dimension of solidarity sharing that will allow humanity to build the future of human life on the planet through new ethical and educational assumptions.

Geoethics (Peppoloni & Di Capua, 2022), developed with the aim of redefining our relationship as humans with the Earth system, proposes a reference framework of principles and values, as well as methods and practices to face the current global challenges, which can be shared at the transnational level. Such a reference framework can accompany the ethical regeneration necessary to direct humanity towards new trajectories of progress.

The ecological, social and international relations crisis requires a renewed awareness in the human abilities to create new models of interaction between humans and between humanity and nature. This implies a radical analysis of what it actually means to mediate between different interests and needs, without this leading to the creation of unbalanced relationships, guided exclusively by the exercise of dominion by the strongest actors.

## 2.3   Geoethics to Realize an Ecological Humanism

Geoethics aims to promote an analytical, critical and scientifically founded attitude towards issues concerning the interaction between the human being and the natural environment, defining cultural categories, ideal principles and behavioural values based on scientific experience and knowledge (Peppoloni et al., 2019; Peppoloni & Di Capua, 2020, 2021a, 2021b, 2022), believing that these can help guide human beings towards more responsible and ecologically oriented individual and social choices. Transferring this attitude to society, understood in all its components (from political decision-makers, to legislators, technicians, the mass media, and citizens) means to favour a cultural change that is prodromal to the modification of the current operational paradigms of the globalized society.

Geoethics has been defined as:

> research and reflection on the values which underpin appropriate behaviours and practices, wherever human activities interact with the Earth system. Geoethics deals with the ethical, social and cultural implications of geoscience knowledge, education, research, practice and communication and with the social role and responsibilities of geoscientists (Peppoloni & Di Capua, 2015, 2021b, 2022; Peppoloni et al., 2019).

This definition outlines the object of geoethical reflection, the perimeter of its analysis and practice, the need to preliminarily identify the values on which to base a more responsible relationship with the Earth, intended as a system. Furthermore, the centrality of geosciences (or Earth sciences) is underlined, as a body of technical-scientific knowledge and practices, which on the one hand allows to understand the ways of functioning of the Earth system and its internal and external interactions; on the other, it contributes to the construction and dissemination of a scientific culture on which to base a more correct management of the interactions between human communities and geo-ecological systems. Therefore, while the first part of the definition outlines the broader vision of geoethical thought and the social-ecological project of geoethics itself, the second part emphasizes the need for geoscientists to be more aware of their social role and of the ethical, social and cultural implications that their activities inevitably entail. In fact, the attitude of those who in the twenty-first century still appeal to an alleged "neutrality" (or indifference, or freedom) of science with respect to the meaning, consequences, and technological applications of the scientific act, seems short-sighted and reductive, even in consideration of the importance that science itself has in supporting the development and progress of humanity. An attitude of prudence and recourse to the analysis of the consequences of the scientific act does not want to curb the potential of science or undermine its freedom of action. Rather, it wants to raise the question of the importance of maintaining a critical and analytical attitude also towards the results of science since they always have a direct and/or indirect impact on society, both in application and cultural terms.

In the specific case of geosciences, the urgency to tackle problems such as pollution, the search for energy and mineral sources, the rise in the average temperature of the planet, the scarcity of fresh water, the increase in the number of disasters triggered by natural and anthropogenic phenomena, the loss of biodiversity, desertification, deforestation especially in tropical areas, as well as the need to improve the adaptive response to climate change, require a new commitment on the part of geoscientists that goes beyond the strictly scientific dimension, going to also involve socio-cultural, political and economic levels. The complexity of the implications of ongoing phenomena requires close cooperation between scientists, scholars and the entire social body since global anthropogenic changes have a generalized impact on the entire system of planetary relations. It is in this perspective that society must renew the conceptual structures and forms that define its identity, programmatic objectives and capacity for agency. In fact, it is not realistic to expect any kind of energy transition without changing the cultural and ethical references of capitalist economy and modern societies, which are reflected in the ecological and sustainable choices made by operators and governments (Peppoloni & Di Capua, 2021a).

In the current historical moment, humanity is at an ethical crossroads: it must choose whether to continue to over-exploit the planet's resources, to neglect the complex physical–chemical-biological-social dynamics of the Earth system and to feed social inequalities or to change, also radically, the dominant economic and financial structures and reform its political systems.

In the environmental field, geosciences are clarifying with ever greater accuracy the profound interconnections and delicate balances of the planet, the irreversibility of some phenomena and therefore the need to adapt to changes (Bohle & Marone, 2022), minimizing their negative effects where possible and reducing the anthropic impact on natural processes and dynamics. In this scenario, even scientists are beginning to promote a collective awareness that encourages the development of global cooperation, capable in turn of addressing the environmental and health problems that affect humanity as a whole, regardless of the political, social, economic and cultural differences of local realities. Far from considering scientists as the custodians of truth and the correct vision of the future, as Bonneuil and Fressoz (2019) also underline, geoethics rather calls upon them to the ethical duty of putting their scientific knowledge at the service of society, and of "*connecting scientific knowledge to the sense of value of society - to what is right, to what is wrong, to what is important*" (Zen, 1993). Furthermore, geoethics underlines the duty of those who practice science to dialogue with non-experts, with ordinary people, or more in general with the final users of their scientific outcomes and results, sharing knowledge and values that lead to consider a common future (Bobrowsky et al., 2017). However, the social commitment of scientists alone is not enough. The general mobilization of the whole of society is needed to influence governments and exercise political action free from restricted temporal logic, capable of planning a future of greater safety, health and sustainability for all.

Each individual, as a "human agent," can benefit from the ethical frame of reference that geoethics proposes as a starting point for orienting their behaviours and actions so that they are more functional to the construction of a new society. In this historical time of enormous danger due to global warming and strong geopolitical tensions, thinking about transforming current societies on the basis on ideals different from the present ones no longer seems like a utopia, but a necessity for humanity's survival.

The project of a society of well-being, of consumption, of wealth has allowed a part of the world to free itself from poverty, to be able to cure diseases, to study and work to improve their material and social conditions. But, unfortunately, this process took place with the exclusion of a large part of humanity and over time resulted in an increase in inequalities and prevarications on a political, economic, social and cultural level, while on the environmental level it produced uncontrolled pollution and a reduction in biodiversity.

From a geoethical perspective, to remedy the *status quo,* it is necessary to increase the environmental and social awareness of individuals since only people who are progressively more informed and aware of the problems affecting the planet will be in a position to decide whether to take responsibility for their choices and actions within ecosystems, combining the pursuit of one's well-being with respect for other human beings, for all other living forms and abiotic elements, with the aim of guaranteeing acceptable living conditions on the whole planet (Peppoloni et al., 2019) and putting an end to that permanent state of warfare within the human species and between it and nature.

Therefore, a distinctive feature of geoethics is that it is centred on the human agent, who is placed at the centre of an ethical reference system, based on three fundamental principles: dignity, freedom and responsibility. In this system, there coexists individual, social and environmental values that derive from those principles (Peppoloni & Di Capua, 2020, 2021a, 2022).

Dignity is expressed by recognizing the intrinsic value of all the elements that make up social-ecological systems and presupposes respect for oneself and for the other from oneself. Therefore, in geoethics, it takes place in the recognition of the value of an entity in the relational architecture that constitutes the ecosystem to which it belongs and by extension to the Earth system.

Freedom is the fundamental requirement, the necessary prerequisite for the human agent to be able to choose and therefore act ethically. It represents the existential condition thanks to which the human being is able to think, process and choose without external constraints that limit their intellectual and decision-making abilities.

Responsibility is the ethical criterion that guides the action of the human agent, who accepts their role within the Earth system and takes into account the possible consequences, even negative ones, of their actions. Human beings, as complex beings, have a responsibility to act in accordance with their complex biological, emotional and rational nature. This implies knowing oneself, being aware of one's potential and limitations, recognizing oneself as moral being and therefore capable of asking ethical questions and making ethical choices.

The principle of responsibility supports human action within the different levels of relationships of the human being, identified in geoethics as domains of experience of the individual: the self, the social group(s) to which they belong (including those professionals), society and future generations, the environment (Bobrowsky et al., 2017; Mogk & Bruckner, 2020; Peppoloni & Di Capua, 2015, 2020, 2022; Peppoloni et al., 2019). These levels become consecutively larger, more complex and intricate as the sphere of agency/relationship of each individual widens.

Geoethics establishes a unity of human action in the various relational domains. Integrity directs coherent behaviours within each of the domains, enriching human existence with meaning: the individual assumes a duty towards him/herself, towards others and progressively towards wider spheres of interaction, up to including the whole Earth system. Through this process, geoethics pushes each individual to feel and become part of the whole, in the recognition of the indissoluble uniqueness of every human being. Perceiving oneself at the centre of oneself is not a selfish attitude, it does not imply domination towards the other from oneself, as in the traditional anthropocentric vision. On the contrary, it is an expression of the fullness of one's authenticity as a human being, well beyond that "epistemic moral anthropocentrism" described by Krebbs (1999) and criticised by Faria and Paez (2014), since it means mainly recognizing an identity of species that is based on the forms in which humans perceive, think and elaborate the world anthropocentrically, which recalls Morin's (2015) concept of "terrestrial identity," through which to give value to what is other than oneself, transcending one's individual perception to embrace a wider one.

Geoethics is characterized as an ethics of individual virtue, in which the *modus vivendi* is articulated through the incorporation into practice of cultural, social and environmental values: among them, sustainability, adaptation, prevention, education to geo-environmental culture and scientific method. Geoethics emphasizes the need for establishing common principles and values of reference for individuals since their social interaction and community organization cannot be ignored. The principle of responsibility is the bridge that binds the cultivation of virtues to the development of a solid moral character on the part of the individual in the vast context of the relationships in which he/she is immersed. Therefore, geoethics is at the same time an ethics of virtue and an ethics of responsibility of the human agent since it implies on the part of the individual, who is striving for virtuous action, the consideration of the consequences of his/her action in a system of complex relationships that will inevitably be modified by his/her choices. The principle of responsibility, on the other hand, qualifies the action of the individual in the social-ecological context of agency. The human being, as a moral subject, is entrusted with the task of addressing the planetary ecological crisis through the awareness of the need to change their ways of relating to the planet, to its resources and other biotic and abiotic entities that constitute it (Peppoloni & Di Capua, 2022).

Sustainable development, a concept originally proposed in reference to the use of natural resources (WCED, 1987), is now used globally also in relation to issues more specifically linked to the human condition, such as the 17 goals for sustainable development included in the "Agenda 2030" of the United Nations.[3]

In the geoethical vision, the concept of sustainable development is redefined to link it, without ambiguity, to a progress that is not only material for humanity. Thus, rather than sustainable development, we believe it is more correct to speak of responsible human development (Peppoloni & Di Capua, 2020).

Human progress is not achieved only in economic terms, but is an articulated set of social, cultural and political products that allow for a better material and spiritual life. Authentic human progress must be achieved with respect for human rights and delicate ecosystemic balances, of which the human being is an integral part. In this perspective, it can give life to an ecological humanism (Peppoloni & Di Capua, 2020, 2022), in which the human being becomes aware on the one hand of his/her complexity, uniqueness and centrality as a modifying agent of the environment, on the other hand also of his/her decentralized elementarity within the natural ecosystem. In this sense, ecological humanism allows to overcome contrasts present in the different positions of environmental ethics regarding the human being-nature nexus (weak and strong anthropocentrism, biocentrism, ecocentrism with its geocentric extension). These positions, despite the progressive attempt to overcome the rigidities present in each of them, however, if taken individually, do not seem to respond to human complexity and overcome an evident dichotomy between humans and nature. Moreover, one cannot fail to highlight that biocentric and ecocentric/geocentric visions refer to concepts developed by the human being, which therefore are inevitably anthropocentric, symbolically representing the human perception of a world of relationships to which it is the human being who attributes an ethical meaning. As humans, an inevitable anthropocentrism of species makes us perceive things in a way and from a position that cannot fail to be anthropocentric, with reference to our relationship with each other from ourselves. Conceiving anthropocentrism in these terms, i.e. referring to the inevitable perception that the human species has of its position on Earth (Viola, 1995), is not in contradiction with respect for nature (Passmore, 1974) and with responsible action towards it, having understood that we are an integral part of it, and that by protecting nature, we are also protecting ourselves. Although environmental ethics has explored possible ways of relating to nature, it has nevertheless synthesized them in positions that have come into conflict with each other (Kopnina et al., 2018; Passmore, 1974). These different positions, despite having opened the discussion on different possibilities of relationship, have in fact created obstacles on the operational level, which if not overcome, today risk slowing down our search for solutions to global problems (Peppoloni & Di Capua, 2021a).

Geoethics therefore proposes a synthetic vision for the human being, which we define ecological humanism, which seeks to grasp the complexity of human feeling and acting and to overcome contrasts, incorporating the concepts of anthropocentrism, biocentrism and ecocentrism/geocentrism into a unitary vision, which saves

---

[3] https://www.un.org/sustainabledevelopment/development-agenda/ (accessed 13 July 2022).

the best insights of the categories of environmental ethics and uses them to develop a new dimension of human relations (Peppoloni & Di Capua, 2021a).

In this view, the human being is characterized as follows:

- intrinsically and perceptively anthropocentric, as he/she cannot escape the human nature, the forms of one's thought, the biological, emotional and rational complexity through which he/she gives meaning to one's perceptions and reflections and constructs the one's vision of existence on the basis of human species peculiarities;
- dynamically anthropogenic, since he/she builds the one's ecological niche to create an operational space, which he/she modifies if necessary to try to improve his/her living conditions.

In the light of geoethical thought, the human being must evolve to become also:

- relationally biocentric (in the recognition of the value of the life of any living being) and ecocentric (in the attitude of respect towards the complexity of the network of ecosystemic interactions and the human role of "*pares inter pares*");
- identitarily geocentric (when he/she develops a sense of supranational belonging to terrestrial citizenship and of stewardship of the Earth system).

## 2.4  An Expanded Definition of Geoethics

The theoretical structure of geoethics has not only the value of intellectual speculation, but is the basis of a pedagogical and political project for the renewal of society, where the human being takes on the responsibility of building the future on ethical assumptions, aspiring to a world more just, supportive and respectful (Peppoloni & Di Capua, 2021a). Therefore, geoethics can be considered a philosophy of complexity (referring to the set of already described domains of geoethics that coexist in human experience), an ethics of individual virtue (as a moral reference system for the training of the individual) and an ethics of social-ecological responsibility (with reference to human action directed towards the social environment and the natural environment). Furthermore, geoethics is also a political ethics, as it solicits the modification of the dominant economic and financial models, which are no longer ecologically sustainable and proposes political decisions that consider the limitation of natural resources on the planet, the irreversible alterations of ecosystems, intra- and intergenerational justice issues relating to access to life-sustaining goods and natural resources. In the geoethical thought, human choices that have an impact on the environment must be scientifically studied, considering the probabilities of possible effects and the uncertainties of forecasting models. Finally, geoethics is also a social ethics since it is built on the civil and environmental commitment of conscious citizens and hopes for an organization of societies that values the richness of diversity, taking advantage of all the speculative and pragmatic opportunities offered by human intelligence to solve common problems.

Geoethical thought has incorporated instances, categories, principles and values already present in the cultural debate: it has developed a theoretical framework that put together those elements, reflections and considerations that animate philosophical, political, sociological, economic and scientific discussions (Peppoloni & Di Capua, 2022). Geoethics looks critically and rationally at issues of global importance (from climate change to the exploitation of resources), attempting to combine causes and effects and paying particular attention to the consequences of activities such as climate geoengineering and deep sea/ocean mining, whose risks have not yet been fully evaluated (Peppoloni & Di Capua, 2020).

Therefore, geoethics can be qualified as (Peppoloni & Di Capua, 2021a):

- *universal and pluralist* (it defines an ethical framework for humanity, in the awareness that the respect of the plurality of visions, approaches, tools is essential to assure dignity to all agents and to guarantee a wide range of opportunities for developing more effective actions to reach common goals).
- *wide* (its reflections cover an extensive variety of themes);
- *multidisciplinary and interdisciplinary* (its approach favours cooperation and overcoming the sectoral languages of the individual disciplines, to reach the intersection and integration of knowledge);
- *synthetic* (it expresses a position of synthesis, definable as ecological humanism, between various existential concepts and different conceptions regarding the nexus between human being and nature);
- *local and global* (its topics of interest concern both the local and regional dimension, and the global one, including the entire Earth system);
- *pedagogical* (it proposes a reference model to cultivate one's ethical dimension, to reach a greater awareness of the value of human identity, not in terms of exercisable power over the other by oneself, but of respect of the dignity of what exists);
- *political* (it criticizes the materialism, egoism and consumerism of capitalism, prefiguring a profound cultural change of economic paradigms and supports the right to knowledge as the foundation of society).

These characteristics define the educational and political goals of a geoethics for society.

Based on these considerations, the definition of geoethics mentioned in a previous section can find a new, broader formulation, which also describes its theoretical structure and operational logic:

> *Geoethics is a field of theoretical and applied ethics focused on studies related to human-Earth system nexus. Geoethics is the research and reflection on principles and values which underpin appropriate behaviours and practices, wherever human activities interact with the Earth system. Geoethics deals with ways of creating a global ethics framework for guiding individual and social human behaviours, while considering human relational domains, plurality of human needs and visions, planetary boundaries and geo-ecological tipping points. Geoethics deals with the ethical, social and cultural implications of geoscience knowledge, education, research, practice and communication and with the social role and responsibilities of geoscientists.*

## 2.5    The Duty to be Human

Geoethics defines at a social and environmental level a logic of relations within the Earth system, created by a human being who acts responsibly towards humanity and the environment, who is aware of his/her role in the Earth system, who wants to assume this role and driven by awareness and a sense of responsibility, feels the duty to act accordingly.

It is evident that initiating the desired change requires a profound cultural and social revolution of the human being. Starting from the educational systems, it will be necessary to begin to form citizens who perceive themselves as terrestrial, inhabitants of the Earth, without this meaning eliminating the cultural specificities of the variegated human mosaic. Furthermore, it will be important to combine the rights of the human being with the mandatory duties of each individual. To this end, an international charter of human duties ("Responsible Human Development Charter" in Peppoloni & Di Capua, 2020) has been proposed, to be understood not as a jurisprudential dictate, but as a means by which every human being can find the best expression of one's humanity in relation to the other from oneself.

In this document, each human being has the duty to:

1. respect the freedom and social, cultural and economic rights of others, rejecting any type of discrimination and recognizing to each one dignity and freedom to develop their own personality, creative potential and talent;
2. develop and exercise responsibility towards themselves, other individuals, the social structures to which he/she belongs, nature in its both animated and inanimate forms, contributing to the realization of inclusive, peaceful, resilient and sustainable human communities;
3. improve one's own knowledge and preparation, within the limits of their possibilities, trying to draw the information useful for their training from sources (institutions, organizations, scientific and technical bodies) that guarantee scientific quality and accuracy;
4. make their knowledge, competence and abilities available to others to contribute to a responsible and ecologically sustainable human development, which also guarantees future generations conditions of well-being, safety and self-determination;
5. ensure inclusivity, equity, solidarity, justice and sustainability in their decisions and actions, taking care that they are based on knowledge that also includes the assessment of possible consequences;
6. cooperate to build and defend socio-economic and political-legal systems that guarantee respect for human rights and the reduction of inequalities, which promote the intellectual, spiritual and material freedom and safety of the individual, without distinction of ethnicity, gender, language, religion, political opinion or economic and social condition;
7. act with prudence and foresight towards the natural environment, considering that one's actions can have repercussions in space and time far beyond what is

foreseeable, due to the intrinsic epistemic uncertainty of the knowledge of social-ecological systems;

8.  contribute, according to one's possibilities, to the identification of a safe and healthy operating space for the human species, by using the resources of nature carefully, respecting their fair distribution and the planetary ecological boundaries;

9.  protect the environment from degradation, from any form of pollution and excessive exploitation, ensuring aesthetic quality of the living spaces and dynamic harmony and balance between their constituent parts.

It is evident that in the geoethical vision the conservation of the Earth's habitability must be configured primarily as an issue of human responsibility: there can be no attention and care towards humanity and the environment without an individual ethical regeneration and a responsibilization of the human being towards the network of relationships in which everyone is immersed.

In this regard, it is important to remember Vaclav Havel's words pronounced on the occasion of his speech to the U.S. Congress in 1990:

> ... the salvation of this human world lies nowhere else than in the human heart, in the human power to reflect, in human modesty, and in human responsibility ... Without a global revolution in the sphere of human consciousness, nothing will change for the better in the sphere of our being as humans, and the catastrophe toward which this world is headed — be it ecological, social, demographic or a general breakdown of civilization — will be unavoidable ... We are still incapable of understanding that the only genuine backbone of all our actions, if they are to be moral, is responsibility ... In other words, we still don't know how to put morality ahead of politics, science, and economics. We are still incapable of understanding that the only genuine core of all our actions – if they are to be moral – is responsibility. Responsibility to something higher than my family, my country, my firm, my success. Responsibility to the order of Being, where all our actions are indelibly recorded and where, and only where, they will be properly judged.[4]

# References

Bobrowsky, P., Cronin, V., Di Capua, G., Kieffer, S., & Peppoloni, S. (2017). The emerging field of geoethics. In L. C. Gundersen (Ed.), *Scientific integrity and ethics: With applications to the geosciences* (pp. 175–212). Special Publication American Geophysical Union, John Wiley and Sons Inc. https://doi.org/10.1002/9781119067825.ch11

Bohle, M., & Marone, E. (2022). Phronesis at the human-earth Nexus: Managed retreat. *Frontiers in Political Sciences, 4*, 819930. https://doi.org/10.3389/fpos.2022.819930

Bollyky, T. J., & Bown, C. P. (2020). The tragedy of vaccine nationalism: Only cooperation can end the pandemic. *Foreign Affairs, 99*(5), 96–108.

Bonneuil, C., & Fressoz, J. -B. (2019). *La Terra, la storia e noi – L'evento Antropocene*. Trad.: A. Accattoli e A. Grechi. Istituto della Enciclopedia Italiana (in Italian). Translation from French of "L'Evénement Anthropocène: La Terre, l'histoire et nous", 2013, Éditions du Seuil.

---

[4] https://www.vhlf.org/havel-quotes/speech-to-the-u-s-congress/ (accessed 13 July 2022).

Callaghan, M., Schleussner, C.-F., Nath, S., Lejeune, Q., Knutson, T. R., et al. (2021). Machine-learning-based evidence and attribution mapping of 100,000 climate impact studies. *Nature Climate Change*. https://doi.org/10.1038/s41558-021-01168-6

Capra, F. (1975). *The Tao of physics: An exploration of the parallels between modern physics and eastern mysticism*. Shambhala. ISBN O-877730776.

Cervato, C., & Frodeman, R. (2012). The significance of geologic time: Cultural, educational, and economic frameworks. In K. A. Kastens, & C. A. Manduca (Eds.), *Earth and mind II: A synthesis of research on thinking and learning in the geosciences*. Geological Society of America, Special Papers 486. https://doi.org/10.1130/2012.2486(03)

Denevan, W. M. (1992). The pristine myth: The landscape of the Americas in 1492. *Annals of the Association of American Geographers, 82*(3), 369–385. https://doi.org/10.1111/j.1467-8306.1992.tb01965.x

Denevan, W. M. (2011). The "Pristine Myth" revisited. *Geographical Review, 101*(4), 576–591. https://doi.org/10.1111/j.1931-0846.2011.00118.x

Ellis, E. C., Gauthier, N., Goldewijk, K. K., Bird, R. B., Boivin, N., et al. (2021). People have shaped most of terrestrial nature for at least 12,000 years. *PNAS, 118*(17), e2023483118. https://doi.org/10.1073/pnas.2023483118

Faria, C., & Paez, E. (2014). Anthropocentrism and speciesism: Conceptual and normative issues. *Revista De Bioetica y Derecho, 32*, 95–103. https://doi.org/10.4321/S1886-58872014000300009

Ferrarotti, F. (2009). *Il senso del luogo*. Armando editore (in Italian).

Fletcher, M.-S., Hamilton, R., Dressler, W., & Palmer, L. (2021). Indigenous knowledge and the shackles of wilderness. *PNAS, 118*(40), e2022218118. https://doi.org/10.1073/pnas.2022218118

Frodeman, R. (1995). Geological reasoning: Geology as an interpretive and historical science. *Geological Society of America Bulletin, 107*(8), 960–968. https://doi.org/10.1130/0016-7606(1995)107%3C0960:GRGAAI%3E2.3.CO;2

Frodeman, R. (2014). Hermeneutics in the field: The philosophy of geology. In B. Babich, & D. Ginev (Eds.), *The multidimensionality of hermeneutic phenomenology. Contributions to phenomenology* (Vol. 70, pp. 69–79). Springer, Cham. https://doi.org/10.1007/978-3-319-01707-5_5

Frodeman, R. (2023). *Geology and knowledge culture*. This volume.

Head, M. J., Steffen, W., Fagerlind, D., Waters, C. N., & Poirier, C., et al. (2021). The great acceleration is real and provides a quantitative basis for the proposed Anthropocene Series/Epoch. *Episodes*, online first. https://doi.org/10.18814/epiiugs/2021/021031

IPCC. (2021). Climate change 2021: The physical science basis. In V. Masson-Delmotte, P. Zhai, A. Pirani, S. L. Connors, & C. Péan, et al. (Eds.), Contribution of working Group I to the sixth assessment report of the intergovernmental panel on climate change. https://www.ipcc.ch/report/ar6/wg1/downloads/report/IPCC_AR6_WGI_SPM_final.pdf.

Jonas, H. (1979). Das Prinzip Verantwortung: Versuch einer Ethik für die technologische Zivilisation. Suhrkamp, Frankfurt/M. The Imperative of Responsibility: In Search of Ethics for the Technological Age (translation of Das Prinzip Verantwortung) trans. Hans Jonas and David Herr (1979). ISBN 0-226-40597-4 (University of Chicago Press, 1984), ISBN 0-226-40596-6.

Jouffray, J.-B., Blasiak, R., Norström, A. V., Österblom, H., & Nyström, M. (2020). The blue acceleration: The trajectory of human expansion into the ocean. *Perspective, 2*, 43–54. https://doi.org/10.1016/j.oneear.2019.12.016

Kopnina, H., Washington, H., Taylor, B., & Piccolo, J. J. (2018). Anthropocentrism: More than just a misunderstood problem. *Journal of Agricultural and Environmental Ethics, 31*, 109–127. https://doi.org/10.1007/s10806-018-9711-1

Krebs, A. (1999). Ethics of nature: A map. *De Gruyter*. https://doi.org/10.1515/9783110802832

Lenton, T. M., Rockström, J., Gaffney, O., Rahmstorf, S., Richardson, K., et al. (2019). Climate tipping points—too risky to bet against. *Nature, 575*, 592–595. https://doi.org/10.1038/d41586-019-03595-0

Meneganzin, A., Pievani, T., & Caserini, S. (2020). Anthropogenic climate change as a monumental niche construction process: Background and philosophical aspects. *Biology & Philosophy, 35*, 38. https://doi.org/10.1007/s10539-020-09754-2

Mogk, D. W., & Bruckner, M. Z. (2020). Geoethics training in the Earth and environmental sciences. *Nature Reviews Earth & Environment, 1*, 81–83. https://doi.org/10.1038/s43017-020-0024-3

Morin, E. (2015). *Insegnare a vivere – Manifesto per cambiare l'educazione*. Raffaello Cortina Editore (in Italian). Translation from French of "Enseigner à vivre: Manifeste pour changer l'éducation", 2014, Actes sud.

Morin, E. (2020). *Cambiamo strada – Le 15 lezioni del Coronavirus*. Raffaello Cortina Editore (in Italian). Translation from French of "Changeons de voie: Les leçons du coronavirus", 2020, Éditions Denoël.

Oreskes N., & Conway E. M. (2019). *Mercanti di dubbi - Come un manipolo di scienziati ha nascosto la verità, dal fumo al riscaldamento globale*. Edizioni Ambiente.

Passmore, J. A. (1974). *Man's responsibility for nature: Ecological problems and western traditions*. Duckworth.

Peppoloni, S., & Di Capua, G. (2015). The Meaning of geoethics. In M. Wyss, & S. Peppoloni (Eds.), *Geoethics: Ethical challenges and case studies in Earth Science* (pp. 3–14). Elsevier. https://doi.org/10.1016/B978-0-12-799935-7.00001-0

Peppoloni, S., Bilham, N., & Di Capua, G. (2019). Contemporary geoethics within geosciences. In M. Bohle (Ed.), *Exploring geoethics: Ethical implications, societal contexts, and professional obligations of the geosciences* (pp. 25–70). Palgrave Macmillan. https://doi.org/10.1007/978-3-030-12010-8_2

Peppoloni, S., & Di Capua, G. (2020). Geoethics as global ethics to face grand challenges for humanity. In G. Di Capua, P. T. Bobrowsky, S. W. Kieffer, & C. Palinkas (Eds.), *Geoethics: Status and future perspectives* (pp. 13–29). Geological Society of London, Special Publications 508. https://doi.org/10.1144/SP508-2020-146

Peppoloni, S., & Di Capua, G. (2021a). Geoethics to start up a pedagogical and political path towards future sustainable societies. *Sustainability, 13*(18), 10024. https://doi.org/10.3390/su131810024

Peppoloni, S., & Di Capua, G. (2021b). Current definition and vision of geoethics. In M. Bohle, & E. Marone (Eds.), *Geo-societal narratives—contextualising geosciences* (pp. 17–28). Palgrave Macmillan, Cham. https://doi.org/10.1007/978-3-030-79028-8_2

Peppoloni, S., & Di Capua G. (2022). *Geoethics: Manifesto for an ethics of responsibility towards the Earth* (pp. XII+123). Springer, Cham, ISBN 978-3030980436. https://doi.org/10.1007/978-3-030-98044-3

Ripple, W. J., Wolf, C., Newsome, T. M., Barnard, P., Moomaw, W. R., & 11,258 scientist signatories from 153 countries (2020). World scientists' warning of a climate emergency. *BioScience, 70*(1), 8–12. https://doi.org/10.1093/biosci/biz088

Ripple, W. J., Wolf, C., Newsome, T. M., Gregg, J. W., Lenton, T. M., et al. (2021). World scientists' warning of a climate emergency 2021. *BioScience, 71*(9), 894–898. https://doi.org/10.1093/biosci/biab079

Rockström, J., Steffen, W., Noone, K., Persson, Å., Chapin, F. S., III., et al. (2009). A safe operating space for humanity. *Nature, 461*(7263), 472–475. https://doi.org/10.1038/461472a

Serres, M. (2019). *Il contratto naturale*. Feltrinelli, Milano (in Italian). Translation from French of "Le Contrat Naturel", 1990, Éditions François Bourin.

Steffen, W., Richardson, K., Rockström, J., Cornell, S. E., Fetzer, I., et al. (2015). Planetary boundaries: Guiding human development on a changing planet. *Science, 347*(6223), 1259855–1259855. https://doi.org/10.1126/science.1259855

Steffen, W., Rockström, J., Richardson, K., Lenton, T. M., Folke, C., et al. (2018). Trajectories of the earth system in the anthropocene. *PNAS, 115*(33), 8252–8259. https://doi.org/10.1073/pnas.1810141115

Stewart, I. S. (2022). Integrating 'The triangle of geography, geology, and geophysics' into sustainable development. *Jordan Journal of Earth and Environmental Sciences, 13*(2), 97–104.

The White House. (1965). *Restoring the quality of our environment—report of the environmental pollution panel*. President's Science Advisory Committee, U.S. Government Printing Office.

Viola, F. (1995). *Stato e Natura*. Edizioni Anabasi SPA, Milano.

WCED. (1987). *World commission on environment and development: Our common future*. Oxford University Press.

Wuebbles, D. J. (2020). Ethics in climate change: a climate scientist's perspective. In G. Di Capua, P. T. Bobrowsky, S. W. Kieffer, & C. Palinkas (Eds.), *Geoethics: Status and future perspectives* (pp. 285–296). Geological Society of London, Special Publications 508. https://doi.org/10.1144/SP508-2020-17

Zen, E. -A. (1993). The citizen-geologist. *GSA Today, 3*, 2–3.

Zen, E. -A. (2018). What is deep time and why should anyone care? *Journal of Geoscience Education, 49*(1), 5–9. https://doi.org/10.5408/1089-9995-49.1.5

# Chapter 3
# The Sustainable and the Unsustainable: What is a Habitable Planet?

Tanella Boni

**Abstract** How can we satisfy the basic needs of a planet that, on the one hand, is becoming increasingly hungry while, on the other, food and natural resources are being wasted? By using an ethical approach—that of virtues—this text intends to show that sustainability concerns the desire to live as a free and responsible human in a society of cultural diversity where living together is possible taking into account social laws and international rules. The sustainable is also about "repairing" the broken bonds between humans, the earth, and all living things. However, in African countries, the desire for sustainability is today hampered by a multiplicity of new ruptures (after those of colonization) instead of reasonable international cooperation and peaceful relations between communities called to live together. It is a question of flushing out the unsustainable in order to think about the field of sustainability that remains to be defined. For the unsustainable is what destroys the bonds between humans and other living things.

**Keywords** Sustainable · Unsustainable · Ethics of virtues · Responsibility · Habitability

## 3.1 Introduction

First of all, I need to clarify in a few words from where I stand. I am neither a geologist, nor a geographer, nor an economist, nor a political scientist, nor a biologist, nor an expert in any science that comes to mind when proposing solutions for the survival of humanity. I have learned to ask myself a few questions about humans who behave as dominators and predators among other living things that populate the earth. I also sometimes marvel at the beauty of the world and tell it in poetry.

Because not a single day goes by without more or less alarming speeches from scientists, from all disciplines, but also from activists—reaching me, about the state of the planet, I can easily imagine that these are not just speeches, but warnings about

T. Boni (✉)
Université Félix Houphouët-Boigny, Abidjan, Côte d'Ivoire
e-mail: sysson2006@yahoo.fr

the catastrophic consequences of climate change that has already begun. Indeed, we have been experiencing them for years: they are hurricanes or storms followed by floods in many parts of the world; they are also droughts that persist with periods of food insecurity; it is the melting of glaciers in the poles and the threat of rising waters. Everything happens as if the clock of planet Earth has gone out of whack. Water, wind, fire, earth overflow their sweet *nature*, as if, at times, having become uncontrollable, they no longer had limits. But the question of limits is one of those that leads us to the ethical concept of sustainability (which limits to growth? what are the limits to technique? how to build a common world?). The world has undoubtedly lost its ecological, but also economic, political, and socio-cultural landmarks. How could it retrieve them? The frequency of exceptional phenomena shows how the exceptional is no longer what it used to be, it acclimatizes, and humans adapt as best they can. From now on, *nature*, once silent or mysterious, transgresses its limits, it bursts into our lives and modes of habitation that it transforms. It "remembers us," said Michel Serres, in the *Contrat Naturel* (Serres, 1991). So, at the crossroads of philosophy and poetry, I see that the conditions of the dwelling of the world are becoming more and more difficult. Isn't that evidence? How can we find a sustainable solution, for the present and for the future, to what seems so obvious to all?

The question, although naïve, leads us to the heart of the global challenges facing humanity today. It could be formulated as follows: in what world have we fallen? But is it enough to be aware of global threats to find satisfactory solutions to local problems? And does the resolution of these contribute to the well-being of humans in the countries and territories where they live? Perhaps we need to think about why nature is not what it used to be. It is no longer local, it becomes global. It must be conceived as the whole of planet Earth and conceived in relation to the history of *civilization* and *development*. However, civilization and development rely heavily on the fabrication of technical tools that transform the world and ways of life. The word civilization, used in the singular, was, during the centuries of great explorations and so-called civilizing missions, from the 18th to the beginning of the twentieth century, one of the concepts justifying the enterprise of colonization of Africa by European countries.

Doesn't this concept already concern the separation of humans and *nature*, the set of ecosystems that bind the living and allow them to keep alive, feed, and die? Civilized is what ceases to be wild, it is believed. What if wildlife helped us to recharge our batteries, to renew ties that we had lost, broken links with nature? But, over time, does it retain its naturalness? We know how much the impact of human activities transforms nature. It is here that an ethics of the earth (Peppoloni & Di Capua, 2022) seems necessary to me.

As for *development*, is it not a vision of well-being in the present that is based on industrial production (no longer on artisanal manufacturing) as well as on the rational organization of national and transnational economies? But who organizes the development of a country? Who helps, who lends the means since finance is at the heart of all development? For almost a century, this concept has contributed to creating imbalances between the regions of the world (the North and the South, for example). It maintains inequalities between rich and poor, while relying heavily on the exploitation of natural resources and the use of fossil fuels.

Thus, the notion of habitation that I put forward includes all these questions that concern the place of the human on earth among other living beings; it takes into account how to use natural resources and non-renewable energies; the production and consumption of raw materials and processed products. Now, is not to inhabit also to weave links between humans and non-humans on the one hand and, on the other, between different localities and the whole world? To understand the notion of "earth" we need to think about the interactions between nature and history, culture, economy, and politics.

Before I develop my point of view, I dare to make a comparison here. Whatever the values to propose, we find ourselves today in a real world comparable to the fictional world created by Lewis Carroll for Alice Lidell (Carroll, 2006). But this world into which we have "fallen," slowly but surely, as if something had escaped us, is far from being "wonderful."

It is our responsibility that is at stake, as is our freedom and dignity (Peppoloni & Di Capua, 2020). However, what are we responsible for when everything seems to escape us, when we have no *control over* anything? And why, in some parts of the world, are humans *crushed* by events of unprecedented violence that they neither wanted nor have foreseen? What happens to them disrupts the order of their world in which they feel lost. Only, astonishment and courage accompany them at every moment, and this keeps them alive. Thus, the researcher of an ethic for a habitable world that I learn to be resembles the character of Alice who does not know who she will meet and when. She does not know the places she crosses. She does not know what other side of the world she will discover. She cannot predict what will happen, since she is walking in an unpredictable world. However, she hopes that by continuing to walk, she will get somewhere. This *hope* is not an expectation during which she has her arms crossed. She knows she has to act. She can only rely on her own strengths and the choices she will have to make. She gives herself the right to ask questions to all the living people she meets on her way, who also give themselves the right to speak. It is in this dialogue that they understand the rules or nonsense of their world.

Thus, the ethics that we seek to highlight has something to do with virtue in the Aristotelian sense of the term, where one acts with caution, according to the middle ground, out of excesses and defects, being excellent in the field of action. Indeed, if we wish to act in the interest of humans and all living on a common planet and in a common world, we must reflect and deliberate on the means that depend on us, the means within our reach and not those that are hypothetical or that remain inaccessible. Yet, being humans of goodwill who care about the other, about everyone else, is something to be learned.

## 3.2   The Sustainable: Translation and Limitations of the Brundtland Report

Let us first recall some proposals made in the twentieth century, in which the word sustainable appears to describe the development that had already been going on for many decades. One example is the *Brundtland Report* (1987)[1] entitled in French *Notre avenir à tous*, used at the Earth Summit in Rio, 1992. The title in English, *Our Common Future*, seems to me to explicitly designate the object of the debate: the community of destiny of all living beings. More than thirty years ago, in the face of the looming ecological crisis, this report proposed common rules for a habitable world. The term "sustainable development," a key word that appears for the first time in this report, indicates that it is not a question of abandoning the idea of development but, on the contrary, of strengthening it while changing its paradigm in order to give it the chance of being applicable everywhere in the world. The essence of the proposals for a sustainable world concerns the needs of man to be satisfied in the present without mortgaging the chances of survival of future generations. The report mentions that the two sciences dealing with sustainability are linked to each other: "Ecology and the economy are indeed closely linked—increasingly, by the way—at the local, regional, national and global levels: it is an inextricable web of causes and effects." Thus, the ecological crisis refers not only to the rupture of the link between man and nature but also between different regions of the world and particularly between North and South.

But what is sustainability? The term "sustainable" indicates that sustainability is the just and ethically acceptable. Thinkers in the French-speaking world first face a problem of terminology. How can we "apply" ideas, however generous they may be, if they are not accepted because they are misunderstood because of inadequate translation? A language is the expression of a culture, a way of thinking, and expressing oneself. How to translate a vision of development that seems to be the opposite of the ways of exploiting the land, producing, marketing, consuming, which had been going on for a long time? In addition, is it certain that "development" is understood in parts of the world in which the word has no equivalent in local languages? As Jean-Philippe Pierron says: "The translation of the term sustainable development involves much more than a technical problem, as the supremacy of English, which has become the language of exchanges, would suggest. It reveals that an approach in technical terms ignores that a language is a living environment, and that the translation of such an expression must acclimatize to specific cultural baths" (Pierron, 2009, p. 34). In each locality, every culture has its own way of conceiving its relationship with *the living environment* in which humans have been *immersed* since birth. In such an environment, the word development might not have any meaning, because the links are woven differently with the other human, with the living and the non-living. There is already a *common* life, more or less ordered, more or less in *symbiosis* which, sometimes, loses meaning or deteriorates because of unwanted

---

[1] https://sustainabledevelopment.un.org/content/documents/5987our-common-future.pdf (accessed 21 September 2022).

events—colonization, war, or the introduction of new techniques or any other disruptive element. Words are understood and thought of differently from one culture to another. That is why using unknown words to designate what to do or not to do, to *improve* the lives of everyone, is a risk. Is this risk good to take?

It is therefore not the term "sustainable development" that poses a problem from the point of view of translation, but "development," a word whose misunderstanding continues to wreak havoc in non-Western countries that have become independent after a long period of colonization. These are countries that live at the crossroads of languages and experiences, such as Côte d'Ivoire, colonized, then independent for more than sixty years and still looking for its own way. The inhabitants of such parts of the world therefore know that what is profitable for others is not always good for them since it is likely to impoverish their lives, if they are not careful. Yet such countries are "developing." But what kind of "development" is this?

Moreover, what does not make it easier for French-speaking researchers is that the expression *sustainable development* has been translated into French by two different words: *durable* and *soutenable*. If sustainable was translated as *soutenable* in Quebec (Canada), the expression *sustainable development* has imposed itself in a large number of French-speaking countries. Should *durabilité* and sustainability be used interchangeably? *Sustainability* includes the idea of long time, beyond the imperatives or emergencies of the present time. But sustainability, without particularly insisting on duration, emphasizes the ethical aspect of things, which preserves the good or well-being of humans and all living things. The aspiration to a good, just, and happy life is at the heart of the idea of sustainability, as is the question of human's place in society and its relationship with nature. In my opinion, the two translations complement each other.

In the 1990s, other debates emerged around the concept of sustainability, especially among economists. Because the strength or weakness of sustainability is relative to what to pass on to future generations, what kind of capital should be left as a legacy? And how? During a discussion on sustainability, published online in 2019, in the journal *Sustainable Development*, Valérie Boivert, Leslie Carnoye and Rémy Petitimbert recall: "… the call to implement sustainable development led to the definition of a sustainability criterion, which was later described as weak by its detractors. To ensure sustainability, a constant capital stock should be passed on to future generations, allowing them at least to maintain an overall level of utility per capita constant. The capital stock in question would encompass natural capital and manufactured capital. Such a representation suggests that the degradation of nature as capital is not a problem in itself, if it can be compensated in its ability to produce utility by various artifacts." (Boisvert et al., 2019).

In 2001, UNESCO, in its *Declaration on Cultural Diversity*, proposed a definition of sustainability. Indeed, it has a role to play in the prevention and resolution of conflicts: "Sustainability can be defined as a criterion of the chances of long-term survival of any desirable human adventure. Therefore, sustainability is the ability to replicate and revitalise essential human resources in the context of new forms of global integration and new opportunities for intercultural dialogue." (UNESCO, 2001, p. 12).

But let us return to the *Brundtland Report*, which has been cited for more than thirty years. Since that time, many other critical texts and discourses have followed; international summits are held periodically to provide food for thought and help with decision-making. Today, it is important to highlight facts about primary human needs. These facts remind us of the urgency of having to act without further delay.

## 3.3   Feeding and Caring for Yourself: Sustainability Criteria

Living—in the biological sense—is in danger, on the one hand because food insecurity is a reality, on the other hand because health, despite all scientific, medical, and technological discoveries, is still threatened by new viruses, as evidenced by the current pandemic. Moreover, malaria is still not eradicated on the African continent. However, prevention and treatment of health problems can be done in a rational way. Wangari Maathai, Nobel Peace Prize laureate in 2004, wonders in *A Challenge for Africa*: "How is it that malaria prevention and treatment is not a priority for African governments? Do we still need to convince a government or a single individual in Africa to protect children from preventable diseases? Why are people not adopting sustainable and effective measures to deal with this disease?" (Maathaï, 2010, p. 83).

To chronically have hunger or sickness without the means to heal oneself are vital, existential experiences that highlight the tragic dimension of the challenges ahead. Food insecurity does not exist everywhere. But malnutrition looms over many children around the world, while, on the other hand, a poor distribution of goods shows how surpluses of agricultural and industrial production are wasted and sometimes thrown away. In some parts of Africa, people are starving. We eat badly. We live below the poverty line. Life expectancy is falling. Maathai, wrote: "Between the last decade of the twentieth century and the first years of the twenty-first century, some African economies began a shift towards growth. But by 2001, the number of Africans living in extreme poverty had nearly doubled, from 164 to 316 million in two decades." (Maathaï, 2010, p. 69). Africans must protect the planet while being able to feed themselves decently, says Maathai, who founded the Green Belt Movement in Kenya in 1977. She was called Mama Miti, the mother of trees. Today, other activists and scientists such as Hindou Oumarou Ibrahim, a geographer, are fighting climate change in the Sahel and defending the rights of Chad's Mbororo Peuhl community. But hunger continues to strike, linked to climate change. In 2021, in the south of Madagascar, because of the drought, the situation was disastrous. Only cacti still resist the lack of water and extreme temperatures. They serve as a means of subsistence for humans who have no other choice. In the Horn of Africa, desert locust clouds invade fields and destroy crops and large tracts of vegetation. Meanwhile, the mice, equally hungry, attack the granaries where the grain is kept. Thus, humans, insects, animals, plants, and earths, linked to each other, survive on the same territory, provided that this fragile balance is not disturbed by excess precipitation or heat. Because the intrusion of climate disasters disrupts everything in their path. It is not enough to foresee them. Ways must be found to re-establish weakened or broken ties. What to do? First, to understand the role of history in telling the metamorphoses of nature and the violence it undergoes.

## 3.4  From Nature to Earth: The Role of History

If we live in a world threatened from all sides, it depends on what we have done with our history, our encounters, our relationships. The human, even if he is born alone, even if he lives alone and dies alone, is a being of relationship. Taking into account the recent history of humanity, the growing concerns are not only a matter of technical and scientific progress, nor of development as such, but of the selected types of development that have fractured the relationship between humans and humans, and humans and nature.

Let us recall here the central place occupied by the idea of nature in Western philosophy from the pre-Socratics (sixth and fifth century BC: Thales, Anaximander, Anaximenes, Heraclitus, Empedocles…) to Descartes and beyond. According to Pierre Hadot, conceptions of nature in philosophy and literature (especially in poetry) take two directions (Hadot, 2004). The first one focuses on respect for nature; the second one relates to the control and transformation of nature. Pre-Socratics considered natural elements, water, air, fire, earth, as the explanatory principles of the world. "Nature loves to veil itself" says Hadot (2004) paraphrasing one of Heraclitus' fragments. Here, nature, seen as mysterious, veiled, and silent, is respected and taken care of as an invaluable treasure. This path seems to have been little followed in the Western world, except perhaps among poets such as Goethe, Novalis, Hölderlin, Baudelaire, Rimbaud. Yet, if we look closely, it was the dominant path in other cultures before the encounter with the Western colonizers. We find traces of this thought at the beginning of the twenty-first century, in Edouard Glissant, when he speaks of Rapa Nui, Easter Island, a distant island lost in the middle of the Pacific: "The magnetic force of the earth is to protect those who come and understand, and those who come and put together, without the absolute being absolutely lost, without also the water below being dried up or soiled, without the lost and defeated being marked by their defeat." (Glissant, 2007, p. 60). Thus, the evocation of certain lands makes us dream. Beyond the violent encounters between humans from different backgrounds. We meditate on the meaning of life because land, even far from where we live, is worth listening to (not just being seen). Now, in a work of writing, to bring together the diversity of imaginaries, is it not to act in order to preserve or preserve something? For example, stories, myths, beliefs, link with nature. Indeed, this long history of nature, which remains mysterious, tells, at times, to what extent it does not "speak" but allows humans to feed, clothe, heal, and satisfy most of their basic needs. Because the living is not only an integral part of nature but needs it. Today, the entire planet carries us and supports us, we hear it, we see its actions of brilliance. How to act?

Nevertheless, in most of the developed world, everything happens as if nature had nothing to hide, since, desecrated, its mysteries no longer exist. The Promethean way, the one that proposes, from the seventeenth century, the idea of mastery and possession of nature through technique and increasingly sophisticated technology, is a dominant thought. This thought, anchored to that of progress and then to that of development, seems to have spread to all continents. As Hadot says: "The Promethean attitude, which consists in using technical processes to wrest from Nature its 'secrets' in order to dominate and exploit it, has had a gigantic influence. It spawned our modern civilization and the global rise of science and industry." (Hadot, 2004). However, technical advances and the industrial revolution will favour the conquest of distant countries, the search for raw materials, transatlantic slavery, and colonization. The *triangular trade* between three continents: Europe, Africa, the Americas, and the Caribbean islands shows how the economic globalization that is taking place has incalculable consequences. Slave traders and some large companies (Dutch, Spanish, Portuguese, French, and British) exported manufactured products (weapons, junk, alcohol, fabrics) from Europe to the African coast where they embarked slaves who would have served as working force in the American land and islands where new ways of dwelling appeared: plantations. From now, the world open *to the elsewhere* is violent and unequal. And the inequalities between rich and poor are glaring. The richness of some allows them to dominate others. On the African coasts, some local chiefs profited from the sale of their fellows to European slave traders. And the black slave is much closer to the trading animal than to the human. It is made to work. The other human who treats it inhumanely thinks of it as *ebony wood*. It seems to me that it is in this way that, as early as the fifteenth century, triangular trade and the slave trade created the unfavourable, unbalanced, unsustainable conditions of habitation in a common world. For centuries (including the Age of Enlightenment) confidence in *progress* and *civilization* has not prevented the world from having several categories of humans: on the one hand *savages*, close to nature, living beings who are only body, arms, and legs capable of working, to whom no dignity was recognized. On the other side, those of *civilization*, who exploited for all intents and purposes everything that could make their lives enjoyable while increasing their wealth (Sala-Molins, 2018).

Thus, the Promethean way of exploiting nature, this way of instrumental rationality which, by wanting to manufacture, produce, innovate, destroys the links between the human and his environment. It has been pursued in violence, in some parts of the world, for centuries. As Thomas Piketty points out: "The distribution in force today between countries of the world as well as within countries bears the deep trace of the slave and colonial heritage. Knowledge of this past is essential to better understand the origins and injustices of the current economic system, but it is not enough as such to define solutions and remedies." (Piketty, 2021, p. 138). It is therefore understandable why, in countries where a large part of the population resulting from slavery is Afro-descendant, *racial* consciousness is a fact that cannot be hidden. The same is true for indigenous peoples who seek their bearings in societies where their cultural rights are barely recognized, where their ways of life are increasingly threatened with extinction. This means that the imaginaries remain marked by history and, whatever the will of each other, the values of dignity, equality, equity, or justice remain relative.

Humans have a fractured consciousness when they feel they share nothing or have nothing in common with their fellow human beings. Some live with the tragic feeling of having no place in a world that is supposed to be their own. Worse, when a hurricane like Katrina occurs suddenly—in 2005 in New Orleans, where the traces of the old plantations are still visible—it is the poorest who are most affected. And, among these poor, there are a lot of blacks. In such a context, perhaps the poor, blacks, and women are the most vulnerable categories when human activities disrupt this nature that is believed to be silent and which, since the twentieth century, has been unleashed more than usual, as if to show that everything is linked: the life of all living people is embarked on the upheavals of the planet itself.

It is therefore not permissible to abandon a planet that, locally, needs care. For it is all the living who, in risk of disappearance, need care at the same time. Who will take care of them? Who will help them regenerate? Perhaps they should be granted rights first? In what sense? It is here that we can mention some environmental philosophers: Aldo Leopold, Arne Næss, Hans Jonas, Michel Serres.

## 3.5   Environmental Ethics: The Place of Nature

Should we be surprised by the ethical, metaphysical, or even mystical discourses of environmental philosophers in the twentieth century? What is the place of sustainability as a humanly desirable life among those who propose new rules for a new object of study: the respectable nature with which it is permissible to *make an alliance*? If thinking about the ecological footprint of human action on the common habitat in order to protect it and conserve it seems to be the starting point of research, the attitude to be adopted towards the object—the living, the earth, the balance between man and nature …—opposes the different currents of this ethic which "aims to determine the conditions under which it is legitimate to extend the community of beings and entities to which men must recognize each other's duties, from the most crude form of animal life to all the ecosystems of our natural environment." (Afeissa, 2007). In this plurality of points of view on human's relationship with nature, it is astonishing that each environmental philosopher insists so much on the novelty of own ethics (Blais & Filion, 2001). What are nature's rights and duties towards it?

Environmental ethics has its cult authors, including Aldo Leopold, who in 1949 published, posthumously, *A Sand County Almanac*. In this work considered founding (although another text preceded it by a century, *Walden; or, Life in the Woods* by Henry David Thoreau, published in 1854), Leopold studied what I would call the *kind of life* of nature. He proposed an ethic of the earth that was meant to be new. Jean-Marie Gustave Le Clézio, in the preface that dates from 1994, speaks of the "revolutionary meaning of *the Almanac*" because, he emphasizes, "what he tells us is clear: that in our world of abundance of goods and impoverishment of life, we can no longer ignore the value of exchange and the need for belonging—this fragile balance that he summarizes in the motive of 'the ethics of the earth' and which will be the concern of the century to come." (Le Clézio, 1994, p. 9). Is it a question of fighting against the impoverishment of life? As an "almanac," this book is first and foremost a reflection on the weather, that of the seasons; he tells us, month by month, what a year is by following the way of life of animals, birds, and the state of the soil, the sun, the earth. Much more than a chronology of events that take place in a locality inhabited by wilderness, the *Almanac* conceives of this land and all its inhabitants as a *community* whose human being is "only one member among others of a biotic team." Human being must respect nature being part of this vast organization that is transformed and complexed over time (Leopold, 2000, p. 259).

If Aldo Leopold's observations lack neither accuracy nor poetry, about, for example, the "international goose trade," from Illinois to the Arctic tundra, we are dealing with a radical environmental ethic that seems to erase in the human being the history of its own culture and that of society in order to allow it to be listening to the mountains, the soil, the waters. Leopold conceives of the disorganization of interlocking life systems as evil because the earth is a living organism that can be in good or poor health. Thus, as we will have understood, the new ethics will have the task of "defining the relationship of human to the earth, to the animals and plants that live on it." One might wonder what is the humanity of the human Leopold is talking about, whether it has a will, whether it thinks, acts, and dreams. Human lives in the good care of the time of the seasons. Thus, would Leopold have locked the human being in a local world, natural more or less mythical, at the time when the concept of development was born, the same year? Is this natural world, left to its own devices, far from any predatory action by human, sustainable for all that? (Wolfgang & Gustavo, 1996, p. 14).

At the birth of environmental ethics, for Leopold the biotope is the place where human being loses its freedom as human being. The world built by human activities, society, and the laws that govern it, politics, culture, everything that constitutes a *humanity* is ignored since only the earth is the centre of the universe.

Other philosophers, such as Arne Næss, will follow the same path. In 1973, he published an article in which he clarified the principles of *deep ecology* (Næss, 1973). In this text, Naess defines this science like no other by the dignity of its object. Thus, deep ecology differs from superficial ecology in that the latter fights against pollution and the depletion of resources with the central objective of "the health and affluence of populations in developed countries." Shallow ecology, he says, is more present and popular. The deep ecology, the principles to which he thinks, states that self-realization merges with the realization of the designs of nature, the true self with which man lives in symbiosis by observing "the equal right for all to live and to flourish." But the analysis of the principles of deep ecology poses difficulties. How to apply the law to a being who has no word, no culture, no will? Will it be able to be aware of this right? The idea of rights and duties towards nature is shared by other philosophers such as Michel Serres, in the *Contrat Naturel* (Serres, 1991).

However, there are many criticisms against the idea of a "natural contract." From this point of view, Luc Ferry, in *The New Ecological Order*, writes, not without irony: "It will be objected, not without reason, that this is a metaphorical fable rather than a rigorous argumentation. It seems very difficult, in fact, to give a proper meaning to the contract proposed by Serres (Hello Mother Nature, I would like to get along with you)" (Ferry, 1992, p 123). Luc Ferry is equally harshly critical of the metaphysical and ecosophical presuppositions of Arne Næss's environmental philosophy. Would he have misunderstood the arguments of those philosophers?

As we know, the other well-known path is that proposed by Hans Jonas: a new ethic based on a heuristic of fear. However, what can we learn from the planetary threat, a consequence of the belief in the limitless power of science and technology? Does not fear scare away? Does it give hope? In *The Principle Responsibility*, he states (Jonas, 2008): "… the promise of modern technology has been reversed into a threat, or that it has indissolubly allied itself with it. It goes beyond the observation of a physical threat. The submission of nature destined for human happiness has brought about by the excesses of its success, which now also extends to the nature of man himself, the greatest challenge that its doing has ever entailed." Therefore, it is a question of proposing ethical standards that play the role of an imperative. Thus, human being's new duty could be thought as *sustainable* in the sense that human actions and their long-term impact will be based on an unwavering principle: responsibility.

Be that as it may, these ethical reflections on the present and the future of the human, who is only one living being among many others, indicate that, by thinking of an ethics of the earth, which forgets neither the question of sustainable development, nor that of the limits of technology, ranging from one science to another, the positions are divergent. The only credible alternative remains that of inter- or transdisciplinarity, which is also that of transculturality. But how can we take into account the concerns of other cultures, other modes of habitation across borders that we think we know? How to integrate global issues into local research? This is one of the questions posed by Workineh Kelbessa, who is concerned with environmental issues in Ethiopia and Africa. Quoting *Aldo Leopold's Almanac*, Kelbessa rightly notes: "Even though Leopold extended ethics to include rivers and soils as well as fauna and flora, his vision is local. His earth ethic does not raise questions about global warming or holes in the ozone layer. Leopold did not question the population explosion or sustainable development, nor, at least, the relationship between rich and poor nations. In terms of ethics, would Leopold have missed the essential, the human-to-human relationship?" (Kelbessa, 2004). Because how to think about the wilderness by staying in the locality, ignoring the immensity of the planet and the diversity of the world? But Leopold lived in the first half of the twentieth century, perhaps he did not really take the measure of the disasters that were coming.

However, Kelbessa believes there is an epistemic justice problem in the field of environmental ethics. He does not have harsh enough words to characterize this attitude as a kind of "racism" (Kelbessa, 2004): "Even today, many writers still do not expect the 'Black Continent', as described by Enlightenment thinkers and anthropologists of the colonial era, to bring ideas about the environment that could help the contemporary world in its quest for environmental solutions." Saying 'Africa' appears, even without saying it, as a kind of black hole of evil. Despite this radical criticism, one cannot ignore the interest of the Western world (now it is also China and South Korea) in the lands, women, and men of Africa. Did not Theodore Monod say in 1950 (Monod, 1950, p. 24): "Africa exists, and it has the right to choose, among the 'various hors d'oeuvres' that we were about to force it to swallow, those that will be truly beneficial to it. Because it can be, it must be admitted with honesty, toxic, for the body and for the soul."

I stress that this is not just an economic or market interest but a general interest. In the search for solutions to global problems, indigenous or *traditional knowledge* also exists in Africa and not only in Asia or America. The perspective envisaged by Workineh Kelbessa is clear: environmental ethics has practical aims, it wants to act, gather knowledge, appeal to those that have existed for millennia and that adapt to the *zeitgeist* (spirit of the age). It is based in particular on the lifestyles of the Oromo pastoralists of Ethiopia, who live by reconciling old knowledge and newcomers from the modern world. His research includes both those on disease (AIDS) and risky behaviours, as well as those on global warming, and many other topics that concern the human world as it is built in contemporary Africa. Perhaps it is in this direction that it is permissible to investigate sustainability as a pooling of particular knowledge to solve, as a whole, a problem of general interest. But this dialogue around a common object is made difficult by a long history of violence and *lack of trust*, between different parts of the world that worry about life and survival on earth. The concept of the earth as a "common home," the Greek idea of *oikia*, comes to mind. Economy and ecology, these two sciences that would concern themselves above all with "home" affairs, are whether allied or in a conflicted relationship. The relationship between the two sciences is clear from the *Brundtland Report*.

## 3.6 Building the Place of the Humans on Earth

After the Glasgow Climate Change Conference (1–13 November 2021), we wonder if the earth is not already lost (Rich, 2019), as it seems difficult for *decision-makers* to agree on concrete actions to be taken to limit the effects of a more serious crisis that is looming. As Isabelle Stengers said a few years ago, if we know that we are going straight into the wall, the "common" nature of this knowledge 'does not reflect the success of a general "awareness." It therefore does not benefit from words, partial knowledge, imaginative creations, multiple convergences that would have resulted in such success (Stengers, 2013, p. 9).

How to make this awareness effective in every individual? For it is first of all the individuals who, locally, are called to resist since their survival is at stake. But individuals do not live alone, they are connected to each other, they have stories, traditions, their own narratives and, at the same time, are integral parts of a global world in the rhythm of which they find themselves embarked. As inhabitants of planet Earth, they suffer the full brunt of the consequences of the disorders of the world of which they do not know all the laws. In such a context, can every human being find the *good* or a *well-being* that is at the same time a *good* for the planet? It is clear that the contribution of a plurality of sciences and practical know-how, coming from diverse backgrounds, will have to be required so that the situation, from a global point of view, can evolve in the right direction.

However, the *common-sense* expression I use here is not the one that Descartes believes is the most shared thing. Because the fundamental question we are facing is simple: how can we do to *be more*? Be more in order to *resist*. Our life can lose its meaning when our body, malnourished or poorly cared for, let us go; when we have trouble breathing, when we are hungry and thirsty. These are concrete experiences of life at the first level, that of primary needs, that lead us on the path of a real awareness of the magnitude of the threats that weigh locally on the survival of individuals, and on humanity in its relationship with all living things. If cultures and beliefs are many and varied, the ways to limit the disasters announced will be just as varied. But they will have to converge in order to meet somewhere. Building a global ethic, suitable for the whole world is at this price: everyone will make his contribution that he will share with his fellow human being for the well-being of all. He will not forget those who do not have articulate speech but who do not cease to remind us who see them, hear, breathe, touch, eat, or imagine: those of the plant kingdom, the animal kingdom, inorganic life, water, air, wind, earth.

However, as Serge Latouche says: "We are not destroying our planet but only our ecosystem, our possibilities to survive it."

Indeed, are we not talking about *survival* first looking around, on a local scale? It is at this level that we can see how *survival* is a word that names the lacks and dissatisfactions that individuals face in their daily lives. Survival is therefore an *unsustainable* life linked, to a large extent, to an imbalance in the distribution of wealth between rich and poor countries. And, in the same society, *having* also made the difference between individuals while establishing an implicit hierarchy between the types of human life. However, each human being is able to give the best of himself, to be virtuous, excellent in action, provided that *sustainable* education and mutual aid systems are put in place and that *favourable* opportunities arise. The main thing sought therefore concerns the ability to choose *to be more*. To be more responsible, freer, more cultured, worthy of being who you are. Provided that the state of *survival* in which such an individual finds himself leaves him the opportunity to find his own resources, to gather his forces in order to *face* what happens to him. Sometimes the choice is not possible. The time of survival is the time of urgency, provided that it is not also the time of ignorance. How to be responsible when the time of choice does not exist? Because making the world habitable and liveable for all is the challenge that must be taken up by every human being in order to prepare and create a sustainable life for future generations.

Giving meaning to the world, out of blind trust in development and growth, is it dreaming? And who would pretend to propose values shared by all regions of the world when many discourses—by scholars or politicians—advocate, during major meetings known to all, solutions to limit future disasters?

It is here that we must think of an ethic of the relationship—not so much by taking the path already marked by Edouard Glissant—but by proposing three orientations that concern: the relationship to the other near, distant or foreign; the relationship to oneself, to one's body, to one's mind; and the relationship to one's living environment.

Perhaps it is necessary to evoke here ethics of life rooted in African cultures. These ethics that continue to resist after colonization. Thus, *Ubuntu*, long before colonial times, established that humans are only human because of the relationship with other humans. In such an ethic, the human being is far from being individualistic or selfish, he takes care of his fellow men, he has a sense of fraternity and welcome. One might think of it as an ethic of resistance in the face of any form of dehumanization. As Munyaradzi Félix Murove shows in an article on Ubuntu (Murove, 2011, p. 45): "As its name suggests, it originated in Africans of Bantu origin as an integral part of their cosmology and the individual ontology that flows from it." The ethics of Ubuntu reflects a way of living together, in courtesy, respect, and benevolence with one's fellow human beings and with nature. But nature is never neutral, it does not stay away from human life. It contributes to the well-being of the human being.

Other ethics, for example in West Africa, can also be mentioned. When the Malian singer Oumou Sangaré sings *Mogoya* (which means humanity in Bambara and Malinké), she evokes a way of being human among all the human beings. Isn't being human the opposite of being barbaric, blood thirsty, *terrorist*? The human being dreams of peace, where it wants to live in peace.

## 3.7  An Ethics of Pacified Relationship

In some cultures, animals are not anonymous living things, nor are plants. We learn to recognize them, to honour their presence. Harvests are also celebrated. This is why floods, deforestation, or droughts are experienced as scourges that strain links forged for centuries with the territories on which humans live. Thus, the earth is the other near, which sees humans born and dying. It is not uncommon to address it, making libations, in the same way that we address a plant to ask permission to pick a few leaves or cut a piece of bark for body care. Emanuele Coccia points out, about plants: "All the objects and tools that surround us come from plants (food, furniture, clothes, fuel, medicines) … our world is a plant fact before being an animal fact." (Coccia, 2016, p. 21).

In colonial literature, these links or addresses to natural elements are seen as traces *of animism*. Yet some prohibitions preserve biodiversity in rural areas where children grow up by internalizing another vision of the plant, the tree, the earth, the water. This way of inhabiting the world is lost in urban areas where *the unbearable* resurfaces in a violent way. The space to breathe is shrinking. Many familiar plants and trees are no longer at hand. Building materials make living spaces stifling. It is in this way, through the degradation of living environments, that the unsustainable settles into the relationship of the human with the environment from which human draws its strength. However, when this relationship—which concerns his various links with everything that makes up his world and nourishes it—becomes fragile, the human being has difficulty in regaining its own dignity, freedom, the sense of fraternity, that of welcome and solidarity. Thus, *sweeping in front of his door*, an ancestral gesture, no longer makes sense. It becomes such hard work that we lose interest in. Who will do this work for us? Sometimes you import the broom and pay a high price, even if you can make it yourself. Consuming local is not obvious, the market system, imposed from the outside and accepted by local decision-makers, thrives because links with oneself have become vulnerable. One example here is rice consumption in West Africa. Cereal consumed daily in many countries, it is imported in very large quantities from Southeast Asia, while many varieties of local rice disappear. That is why, one of the first values to propose in order to resist all kinds of disasters is the ability to have a *well-made head*, according to Montaigne's expression. *To be more* is to have an open mind, it is to refuse to submit to the dictates of ignorance that is confused, in some regions of the world, with the ideology of development which is far from sustainable. Only the awareness *of being more than* we are led to believe, regardless of the cultures we share, allows us to be truly responsible humans. We could talk about education or instruction, but we must go further. Becoming aware of planetary disasters is not enough. The ethics we seek concerns our multiple relationships with the other, all others, living or non-living, and with the earth whose presence we learn to recognize, in all circumstances.

# References

Afeissa, H. -S. (Ed.). (2007). Ethique de l'environnement: Nature, valeur, respect. Vrin, Paris. ISBN 978-711619436.

Blais, F., & Filion, M. (2001). De l'éthique environnementale à l'écologie politique Apories et limites de l'éthique environnementale. *Philosophiques, 28*(2), 255–280. https://doi.org/10.7202/005664ar

Boisvert, V., Carnoye, L., & Petitimbert, R. (2019). La durabilité forte: Enjeux épistémologiques et politiques, de l'économie écologique aux autres sciences sociales. *Développement Durable Et Territoires, 10*(1), 1–16. https://doi.org/10.4000/developpementdurable.13837

Carroll, L. (2006). Les aventures d'Alice au pays des merveilles: De l'autre côté du miroir et de ce qu'Alice y trouva (trans: Parisot H.). Diane de Selliers, Paris.

Coccia, E. (2016). La vie des plantes: Une métaphysique du mélange. Rivages, Paris. ISBN 978-2743638009.

Ferry, L. (1992). Le nouvel ordre écologique: L'arbre, l'animal et l'homme. Editions Grasset, Paris. ISBN 978-2246468110.

Glissant, E. (2007). La terre magnétique: les errances de Rapa Nui, l'île de Pâques. Seuil, Paris. ISBN 978-2020899031., p. 60.

Hadot, P. (2004). Le voile d'Isis: Essay on the idea of nature. Gallimard, Paris, ISBN 978-2070730889.

Jonas, H. (2008). Le Principe responsabilité: Une éthique pour la civilisation technologique. Flammarion. ISBN 978-2081213005.

Kelbessa, W. (2004). La réhabilitation de l'éthique environnementale en Afrique. *Diogène, 207*(3), 20–42. https://doi.org/10.3917/dio.207.0020

Le Clézio, J. -M. G. (1994). Préface. In: Leopold A., Almanach d'un comté des sables: Suivi de Quelques croquis. GF Flammarion, Paris.

Leopold, A. (2000). Almanach d'un comté des sables. GF Flammarion, Paris. ISBN 978-2080710604.

Maathaï, W. (2010). Un défi pour l'Afrique. Héloïse d'Ormesson, Paris, ISBN 978-2350871400.

Monod, T. (1950). Le monde noir. Présence africaine no spécial 8–9, Paris.

Murove, M.F. (2011). Ubuntu. Diogène, 235–236(3–4), 44–59. https://doi.org/10.3917/dio.235.0044

Næss, A. (1973). The shallow and the deep, long-range ecology movements: A summary. *Inquiry, 16*, 95–100. https://iseethics.files.wordpress.com/2013/02/naess-arne-the-shallow-and-the-deep-long-range-ecology-movement.pdf

Peppoloni, S., & Di Capua, G. (2020). Geoethics as global ethics to face grand challenges for humanity. In: G. Di Capua, P. T. Bobrowsky, S. W. Kieffer & C. Palinkas (Eds.), *Geoethics: Status and future perspectives* (Vol. 508, PP. 13–29). Geological Society of London, Special Publications. https://doi.org/10.1144/SP508-2020-146

Peppoloni S., & Di Capua, G. (2022). *Geoethics: Manifesto for an ethics of responsibility towards the earth* (pp. XII+123). Springer, Cham. https://doi.org/10.1007/978-3-030-98044-3

Pierron, J. -P. (2009). Penser le développement durable. Ellipses, Paris, ISBN 978-2729852023.

Piketty, T. (2021). Une brève histoire de l'égalité. Seuil, Paris. ISBN 978-2021485974.

Rich, N. (2019). Perdre la Terre Une histoire de notre temps (Traduit par: Fauquemberg D.). Seuil, Paris. ISBN 978-2021424843.

Sala-Molins, L. (2018). Le Code Noir ou le calvaire de Canaan. PUF, Paris. ISBN 978-2130813323.

Serres, M. (1991). Le Contrat naturel. Bourin Julliard, ISBN 978-2724253726.

Stengers, I. (2013). Au temps des catastrophes: Résister à la barbarie qui s'annonce. La Découverte, Paris. ISBN 978-2707177193.

UNESCO. (2001). Universal declaration on cultural diversity. http://unesdoc.unesco.org/images/0012/001246/124687e.pdf#page=67. Accessed August 26, 2022.

Wolfgang, S. et Gustavo, E. (1996). Des ruines du développement, Montréal, Ecosociété. ISBN 978-2921561051.

# Chapter 4
# Geology and Knowledge Culture

Robert Frodeman

> *The result, therefore, of our present enquiry is, that we find no*
> *vestige of a beginning—no prospect of an end.*
> James Hutton

**Abstract** Since its creation in the late nineteenth century the research university has treated geology as a regional ontology—as one more body of knowledge alongside the other disciplines. The imperatives of the twenty-first century suggest that this needs to change. The purposes of knowledge production reflect the goals of a culture; as those goals change so should the nature of knowledge production. As sustainability becomes the overarching goal of all our efforts our knowledge culture needs to reflect this fact. This implies that geology should become the framework for all knowledge production, facilitating the birth of a new society of maturity and limit.

**Keywords** Interdisciplinarity · Deep time · Critical university studies · Philosophy of geology · Sustainability

## 4.1 Introduction

The modern research university treats geology as a science. Geology forms one more element within the horizontal taxonomy of the disciplines, neither higher nor lower nor more central than any other field. Geology obeys the ontology of the academy, where subjects occupy discrete domains within either the natural sciences, the social sciences, or the arts and humanities.

This view is mistaken about both geology and the nature of knowledge. Of course, geology is a science. But it is also an inter- and transdisciplinary field that overturns the theoretical assumptions of the university. Geology is the domain of deep time, the integrative element of the sciences, and the foundation of a sustainable worldview.

R. Frodeman (✉)
1660 JW Drive, Jackson, WY 83001, USA
e-mail: robert.frodeman@gmail.com

© The Author(s), under exclusive license to Springer Nature Switzerland AG 2023    41
G. Di Capua and L. Oosterbeek (eds.), *Bridges to Global Ethics*,
SpringerBriefs in Geoethics,
https://doi.org/10.1007/978-3-031-22223-8_4

At its furthest extent, geology provides us with the framework of a general theory of limit and a roadmap for restructuring our social norms.

Today we appreciate the permeability of disciplinary boundaries. We understand that disciplines leak into one other. But geology goes further: it exposes the limits of a disciplinary approach to knowledge. As Earth systems science, geology encompasses the other sciences. It frames our lives within new and critically important historical perspectives. Its insights raise pressing social, ethical, and metaphysical issues. Geology has long functioned as helpmate to industrial society, supplying minerals and energy to sustain the status quo. In the twenty-first century its main role should shift to facilitating the birth of a new society of maturity and limit.

## 4.2   Current Efforts

In recent years a group of geologists have sought to draw out the larger implications of geology. They have done so via the concept of geoethics (e.g., Peppoloni & Di Capua, 2015). These thinkers are motivated by the fact that the relation between humans and the planet they inhabit has fundamentally altered. This realization also lies behind other attempts to describe the challenge before us, for instance, in discussions concerning the naming of a new geologic era called the Anthropocene.

The challenge facing these efforts is to generate a conceptual response adequate to the imperatives of a new period in human history. This is no small task. I view this challenge in terms of breaking our addiction to growth and ushering in an age organized around the concept of maturity. The changes this implies would be profound: shrinking the world's population, halting the endless expansion of the economy, tempering our Faustian scientific impulses, setting aside some of our toys, and recognizing the necessity and beauty in limit.

This view presents two challenges to geoethics. First, while Peppoloni and Di Capua emphasize the conceptual breadth of geoethics, the term threatens to limit the influence that geological thinking should have on society. This reflects a bias within culture at large where ethics is the sole philosophic category worth attending to. Just as the science of geology is understood to encompass all of Earth system science (the lithosphere, hydrosphere, biosphere, etc.), a wider notion of geoethics should involve all areas of philosophy (esthetics, metaphysics, political philosophy, etc.), the arts and humanities generally, as well as the policy dimensions of its insights.

Second, the task before us is both larger and more fundamental than simply drawing out the cultural dimensions of the Earth sciences. Geology reveals the theoretical and institutional limitations of the modern knowledge enterprise as it is embodied in our universities. Properly understood, geology offers a pervasive critique of the epistemic status quo. It challenges the way we have defined the knowledge enterprise over the last 150 years. It critiques the aims and structure of the modern research university and the society that it serves.

At its furthest extent, geology should join hands with the more general project of rethinking the nature of modern culture (Frodeman, 2014, 2019).

## 4.3 The Huttonian Revolution

Geology dates from the time of Werner and Hutton at the end of the eighteenth century. Philosophy traces its origins back some 2500 years to the persons of Heraclitus and Parmenides, Socrates, Plato, and Aristotle. If we ask whether there is anything substantive that connects the two subjects the first issue that comes up is the concept of time.

Geology (as opposed to mineralogy) was born out of a novel understanding of time. Werner realized that rock units could be defined in terms of time rather than composition, and Hutton intuited the incredible lengths of time represented in the rock record (Laudan, 1987). Then with the discovery of radioactivity early in the twentieth century we were able to put firm numbers on the immensity of Earth history.

The late eighteenth century also saw philosophy being reshaped by temporal perspectives. In 1797 Schlegel coined the term historicism, and soon thereafter Hegel described different philosophical systems not as a series of rejections but as the progressive development of human consciousness across time. In the early twentieth century Heidegger argued that our assumptions about time fundamentally shape our sense of reality. With little or no attention being paid to geologic time, 20th and now 21st-century cultural studies have been deeply historicist in orientation.

Deep time is not the only geological concept that spans the science-humanities divide. Geology contains a rich set of terms that escape disciplinary control—sedimentation, lithification, and metamorphism, uniformitarianism and catastrophism, erosion and angle of repose. But like space (compare the Copernican Revolution), time defines a basic parameter of existence. While taking no notice of geology, Heidegger's masterwork *Being and Time* (1927) placed time at the center of our understanding of reality. As he states in his preface, time is "the possible horizon for any understanding whatsoever of Being." For instance, the Christian idea of an immortal soul presupposes that reality consists of two parts, one of which exists outside the flow of time.

There has been speculation on the cultural implications of geologic time, most commonly by geologists. The conclusion most often drawn is that placing the last few thousand years of human experience against the immensity of geologic time reduces humanity to insignificance. One can just as easily arrive at the opposite conclusion, that the human enterprise is ennobled by being placed within the framework of this stupendous history. In truth, neither narrative takes us very far. Both offer only rudimentary accounts of the impact of geologic time on our self-understanding. They both reduce a multitude of possible insights to a single narrative.

For instance, geologic time is scalar, and so the lessons that we draw will be different at different time scales. From the point of view of the Pleistocene, human culture has developed in the middle of a thaw. But from the perspective of deeper time, we are still in the middle of an Ice Age: there has been permanent ice on the surface for only 7% of the history of the Earth. The implications of geologic time vary depending on the topic, as the geologic record reveals strange creatures, diverse landscapes, and wide-ranging climates and conditions. For instance, what counts as an exotic species varies by time scale: from the perspective of the last Ice Age 18,000 years ago every species in Yellowstone is an exotic (Pleistocene Yellowstone was buried under an icecap). The rock record also becomes more fragmentary the further back we go even as the planet it reveals becomes weirder.

The points revealed by deep time raise issues that are as much ones of psychology, politics, and culture as of science. Paul Shepard argued that human consciousness needs to be understood against the background of deep time. Not only our body but also our consciousness evolves over time; the modern world of huge cities, artificial light, and constant electronic stimulation has left us permanently off kilter (Shepard, 1982). The awareness of deep time should influence every domain of human and natural history. The crises we face, most notably climate change, demand that we simultaneously think of time in its human and geologic dimensions. This means learning to stretch our awareness across the lags between geologic time and the time scales of everyday existence.

Whether we focus on electronic, human, or geologic scales, time not only entails change; it also implies limit. When Heidegger speaks of human life in terms of being-toward-death (*Sein-zum-Tode*) he is emphasizing that it is through recognizing our finitude that a life of integrity becomes possible. Limits force choices, where we stake a claim and commit to a way forward—or not. Death is the ultimate limit, but life presents us with many other points where something has ended. Ignoring these limits, or pretending that they are not real, is to succumb to a bad infinity—as is our culture's constant demand for new toys.

As Hutton notes, geologic time can seem nearly limitless, stretching back 4.5 billion years and into an indefinite future. But by placing our actions in a larger context geologic time highlights our own particular finitude. The Ogallala aquifer may seem inexhaustible, but we are mining Pleistocene water for the needs of a few decades. A century and a half of burning fossil fuels seems inconsequential until we understand that we are creating an atmosphere last seen in the Miocene. The limits geology points to—of how much carbon and methane we can put into the atmosphere before disaster results, or the point at which continued clearing of the Amazon will turn it from forest to savanna—portend the end of the culture of infinite desire. Geologic knowledge and perspectives imply the need for a new culture of maturity.

## 4.4  Making Deep Time Intuitive

Humans are short-sighted creatures; long-range planning is a rare accomplishment. Even then, what counts as long-range is measured on human scales. Events involving geologic spans of time—the rate at which lost soils are replaced or a degraded ecosystem reconstitutes itself—are essentially discounted to zero.

Making deep time intuitive will require an innovation in human attentiveness. In *The Genealogy of Morals* (1887) Nietzsche asked how a hairless ape managed to become human. He argues that the process required more than intelligence and an opposable thumb; human society also depended on the ability to commit to future outcomes. Humans had to develop the capacity to make and keep promises.

Nietzsche argued that a promise became a commitment through the lessons of pain: suffering the consequences of breaking a promise burned future pledges deep into our soul. This gloomy analysis remains relevant today. Acknowledging the existence of environmental limits depends on the capacity of intuitively grasping long stretches of time, where "now" extends beyond the moment to include decades and centuries into the future. Our halting response to the dangers of climate change marks our disregard of the fact that geologic time is also our time.

Our tepid response to the dangers of climate change suggests that Nietzsche is correct: only widespread suffering will motivate culture-wide transformation. Our lack of action will have consequences on the far side of things as well. Few realize that once the climate has changed the new conditions will be irrevocable on human time scales. Absent breakthrough technologies like carbon capture and sequestration the modified climate will be with us for centuries to come: $CO_2$ remains in the atmosphere on a time scale of centuries.

Nietzsche may be correct about human nature. Nonetheless, it is worth searching for less traumatic ways for extending our temporal horizon. Seeing current events through the lens of deep time should become part of our education from the first years of schooling. I do not mean lessons that focus on representations of geologic time like those that are standard within geology courses (e.g., comparing 4.5 billion years to a calendar year). Rather, we should make geologic time more real through intuitive accounts of one's local surroundings.

I can cite an example from my own work. In 2002, as part of a National Science Foundation-funded project in curriculum development, museum-quality signage was mounted on the outside wall of an elementary school in Boulder, Colorado. The school—Flatirons Elementary—sits at the border of two geologic provinces. The Laramide orogeny rises immediately to the west of the school: the mountains have burst through the sediments of the Cretaceous mid-continental seaway which run for hundreds of miles to the east.

The image below depicts what the area looked like 90 million years ago. Boulder sits at 1650 m/5400 feet of elevation; in the Cretaceous the location was under 760 m/2500 feet of water. The image is attached to a wall facing the school playground, meaning that students are exposed to it daily across their grade school years (Johnson et al., 2005) (Fig. 4.1).

The image comes from a project known as *Ancient Denvers*, a collaborative effort involving geologists and landscape artists funded by the National Science Foundation. The project sought to depict the paleoenvironments of the various strata of the Denver, Colorado area. Scientists collaborated with artists to show landscapes across geologic time, working from the latest science to create accurate and evocative images of the past. The resulting works formed the basis of a 2003 exhibition held at the Denver Museum of Nature and Science.

It is often claimed that geologic spans of time are incomprehensible to a species that lives for less than a century. This sells our imaginative capacities short. Sustained

**Fig. 4.1** 65 × 52 cm image attached to the outside wall of Flatirons Elementary, with the label: "Boulder, Colorado 90 million years ago." *Ancient Denvers*, 2005

exposure to deep time eventually reshapes one's sense of reality. Evidence of this can be found in the experience of geologists who have spent a lifetime in the field. John McPhee, who coined the phrase "deep time," demonstrates the point across a series of books. *In Suspect Terrain* recounts the comments of a geologist concerning the proposed protection of the Boundary Waters Area in Minnesota. While favoring protection, she likens those lakes to the puddles left after a rainstorm. The lakes are the last remnants of the melting ice sheets: "Another five thousand years and there won't be much to fight about," Anita said, with a shrug and a smile. "Most of those Minnesota lakes will probably be as dry as these in Indiana" (McPhee, 1983).

The point is not to dismiss the protection of the area, any more than we would disregard a broken arm because the person will be dead 100 years from now. Geologic time helps us reframe our challenges so that we can be more strategic in our decision-making.

## 4.5 Epistemic Assumptions

There is more to be said on the cultural implications of deep time. In fact, the topic deserves its own policy-oriented research program. But set this to one side, for the point of this essay is to survey the overall significance of geology.

Reflection on geologic time should be complemented by attention to the integrative aspects of geology. Geology (or as it is also termed, Earth system science) unites the other sciences. Geoethics, or perhaps better said geophilosophy, should highlight the preeminent goal of all future social policy—tending to the health of the planet that all life depends upon. Doing so will not only raise questions of ethics, value, and policy. It should also underscore how geology challenges the assumptions of the research university and of contemporary knowledge culture. In contrast to the structure of the contemporary university, knowledge has become hierarchical again. A preeminent value—sustainability—should unite all our epistemic efforts.

Thomas Kuhn argued that most academic work consists of puzzle solving, as researchers strive to make small advances within a disciplinary or sub-disciplinary field. Within these areas there occasionally arise thinkers who challenge the assumptions underlying these research programs. These theorists—the Einsteins or Crick and Watsons of the world—are engaged in radical critique, shifting the paradigm of their field. But Kuhn did not consider the possibility of another level of critique. His paradigm-breakers leave the overall structure and goals of the institution they are housed within intact. Today it is knowledge culture itself that needs a Copernican Revolution.

The modern research university is built on two linked assumptions. First, knowledge is flat: no discipline is viewed as more fundamental than or superior to another. Second, the production of knowledge is an infinite project. There is no end to knowledge production because there is no end to our desires. These two assumptions are so deeply embedded within academic culture and society at large that they are not even subject to debate.

Begin with the first point. Despite increasing attention paid to interdisciplinary approaches, disciplines still dominate the academy. Each operates as a largely separate domain. Clark Kerr, president of the University of California system across the 1950 and 1960s, described the modern university as a "multiversity" serving a vast number of constituencies and interests. The university has no overarching purpose other than the endless pursuit of knowledge. The knowledge it produces has no specific end: it provides a buffet that individuals (or corporations) select from as they see fit.

Compare this with the European medieval university and the American colonial college. Both believed that knowledge had an overall purpose. Knowledge was inherently hierarchical in nature. Individual projects were pursued, and subordinate goals achieved, but there was general agreement about the overall rationale for the institution: education served a religious end.

This was reflected in the structure of these institutions. In the medieval university the division of professors into higher and lower faculties expressed the fact that some types of knowledge were subordinate to others. Within the three higher faculties of medicine, law, and theology formed an ascending order: medicine was concerned with the health of the body, law with the health of the polity, and theology with the health and destiny of our immortal souls. One sign of the non-disciplinary nature of the university was the fact that professors would often move through the different faculties across their career (Clark, 2009).

Similar beliefs characterized the early American college. The senior capstone course in moral philosophy was usually taught by the college president. His role was to pull together the threads of a college education toward overall goals that were both personal and social in nature—one's own salvation and the development of a sense of noblesse oblige, where the fortunate act with generosity toward those less privileged.

The ultimate objective of these institutions was eschatological in nature, the saving of one's own and other's immortal soul. Of course, such a goal today is impossible, at least within public institutions. The question of what constitutes the good life is a private matter, and values are seen as inescapably pluralist in nature. Following social contract theory, politics has been reconstructed to make minimal demands on its citizens, and society now has a libertarian cast.

The restructuring of our knowledge institutions was crucial to the Enlightenment project (and also to the goals of capitalism, which sought profit through innovation). Christian beliefs concerning the *summum bonum* were thrown off as people became free to do as they wish in their lives, subject to minimal conditions. These conditions were codified by John Stuart Mill: people should be free to act however they wish unless their actions caused harm to others.

Few noticed that Mill's argument contained a geological premise. It presupposes the existence of a vast storehouse of resources large enough that their use by one person or group did not affect the prosperity of others. But under conditions of scarcity, one's actions cannot be isolated. The pluralism of contemporary culture, where we treat the existence of irreconcilable differences in life goals as both an inescapable fact and as a virtuous invitation to develop one's individuality, presumes abundance.

The COVID-19 pandemic highlighted the breakdown of Millsian logic: claims that one may choose to not be vaccinated or to wear a mask ignored the fact that these actions inevitably affect others. For scarcity comes in many forms: COVID-19 underlined the scarcity of social space just as the climate crisis demonstrates the lack of sufficient amounts of atmosphere to harmlessly absorb all the carbon dioxide and methane we have been emitting.

Like society, the modern university has been built on a libertarian logic. The smorgasbord approach to knowledge, where its products are treated as a means to whatever ends an individual wants to pursue, assumes a world where we need not consider how inventions or discoveries behave when released within society. The operating assumption, again tacit, is that we can count on all these combinations being benign in their social effects.

Modernity is defined by the development of a libertarian culture whose ever-widening choices are provided by new discoveries in science and technology. Over the last few decades some have predicted the rise of a new, post-modern era. One view sees post-modernity as marking the end of all meta-narratives, those overarching accounts of life that provide a structure for people's beliefs and give meaning to their experience (Lyotard, 1979). The problem with this claim is that humans always reply on some type of meta-narrative, even if it consists of nothing more than the claim that metaphysics is dead and all we have left is physics and our endless desires.

The meta-narrative of modernity has consisted in its belief in progress—the continual satisfaction of our desires through constant innovations in science and technology. Today we are at the cusp of a new meta-narrative where we recognized a common end to society, based not in Christianity or technoscience but in geophilosophy.

## 4.6   Infinite Knowledge

Turn now to the second premise of the modern research university: the production of knowledge as an infinite task. While never stated, much less debated, this is the norm within every discipline. Except for a few holdouts in the humanities who believe in the idea of a *philosophia perennis*, this view is accepted by everyone across the academy.

Few within the academy realize that this assumption is of recent vintage. In the past people were suspicious of *libido sciendi*, the lust to know. This attitude is still visible in the stories we learn as children, of Icarus, Pandora, Faust, and Frankenstein. As Roger Shattuck (Shattuck, 1997) details, the view was prevalent for millennia, only shifting with the advent of modernity. Immanuel Kant summarized the spirit of modernity when he cited Horace's phrase *sapere aude!*—dare to know. In recent years this view has also become prevalent across the humanities, in the rejection of the idea of a canon of works of perennial value.

Given present circumstances, it is worth asking what premises concerning knowledge production best serve the future of humanity. The answer turns on understanding the place of geologic knowledge—or if you prefer, ecological knowledge with the added perspective of deep time—in the theoretical architecture of the university and in society at large. Earth scientific knowledge is not simply another body of knowledge alongside others. This knowledge, and the societal consequences we draw from it, offers us the outline of a new meta-narrative. Society will still pursue myriad ends. But all of these will need to be checked in terms of their sustainability. This fact should affect the nature of knowledge production and lead to the restructuring of the university as well as the society that it serves.

The assumption of infinite knowledge is connected to the flat and regional nature of the disciplines. Restricting every subject to its own region of being—including philosophy and the humanities, which traditionally had sought to offer a view of the whole—has meant that there has been no organized discussion of the overall purpose of our epistemic efforts. Instead, knowledge production, structured as a group of regional ontologies, has treated knowledge as a means—a rational means to private and often irrational ends. The lack of an end in the sense of limiting knowledge production is a consequence of the lack of end in the sense of there being no overall purpose to knowledge.

At the founding of the research university at the end of the nineteenth century this approach was commendable. We had much to learn in terms of basic health and welfare. A radical pushing of all boundaries made sense to, as Bacon put it, "relieve man's estate." An increasingly detailed focus within each of the sciences served us well. The discoveries made lessened many of the burdens that had long tortured humanity. And our technologies were not so advanced as to raise question of their threatening our well-being.

But the function of knowledge changes over time. The projects and attitudes of one period need to be rethought in another. Within society the pursuit of infinite knowledge has been known as progress. That term has largely been defined in terms of material and technological development. About 150 years after the founding of the research university this mission remains the same. No distinction is drawn between the pressing needs that have been addressed (e.g., sanitation, striking advances in medicine, and adequate food production) versus the satisfaction of peripheral desires (larger homes, a new app). Nor do we distinguish between satisfying the urges of those in developed countries, whose basic needs have been largely met, versus the situation in those parts of the world still lacking basic services.

Every culture, past and present, makes epistemic efforts. But only one culture has created a system for the continuous production of knowledge to provide an unending stream of (so-called) improvements in our lives. The rational for these efforts seems self-explanatory. For we all want to continue to grow the economy, conquer disease, and address environmental problems.

To state such goals in a piecemeal fashion, as both researchers and the public do, is to make a point that seems irrefutable. Of course, we wanted vaccines to end the pandemic, cleaner sources of energy, and more efficient transportation. The list is as endless as are our desires. But this is to commit the fallacy of composition, the assumption that when the members of a collection all share a property the collection as a whole possesses that property as well. Our individual desires may make sense (some do not, or are trivial, but let that pass). But what happens when they are aggregated? Academics, housed within disciplines, all pursue knowledge of one type or another. But where does this piecemeal process take us when considered as a whole?

Transhumanism provides an answer to this question. Transhumanists approach the knowledge enterprise as a whole, asking about the overall direction of science and technology. Their conclusion is that science and technology are moving us toward a condition of infinite human power. Transhumanists differ on the particularities of how this process will be achieved—perhaps though the physical and cognitive augmentation of our simian bodies, perhaps through a silicon future as artificial intelligence comes to either serve, blend with, or absorb us. But by whichever means, they view the end result as clear: deification.

Transhumanism is typically dismissed as the obsession of a few oddballs. More accurately, transhumanists have revealed the tacit goal of modern culture. Whether judged in terms of capitalism, or the belief in continual scientific and technological progress, or simply in terms of the nature of human desire, our culture's love of infinity is tacitly transhumanist in orientation. Transhumanists make explicit the logical endpoint of the Enlightenment project (Frodeman, 2019).

Once attuned to this the transhumanist impulse can be seen everywhere. The US National Science Foundation places no limit on its program of scientific and technological advance, just as the US National Institutes of Health hope to overcome every infirmity. The same is true for every other nation's path of research. The only difference between the transhumanists and the rest of us is in the degree of self-awareness of where things are trending. Our epistemic trajectory points toward infinite power; transhumanists have simply made the point explicit.

Transhumanists deserve praise for achieving a global view of our situation. But this clarity raises a new set of questions, the most basic of which is whether the goal of infinite power is a desirable one, or whether like the Sorcerer's Apprentice the process is likely to spin out of control. It is unclear that we are taking the dangers of the continued laissez faire knowledge production seriously enough. The endless pursuit of technoscientific knowledge will lead to any number of improvements. But as our knowledge increases so does our power, which can be used in both beneficial and destructive ways.

Whenever limits have appeared scientific and technological advance has made it possible to transcend these limits. This is why economist Julian Simon called human creativity the ultimate resource. For decades predictions were made concerning peak oil, the point at which petroleum reserves would reach their high point and start their inevitable decline. Then technological advance (directional drilling, hydraulic fracturing aka fracking) resets the entire question.

Technology may or may not come to the rescue to solve today's problems. But even if Simon is correct about our creative abilities, we still face a dilemma. Innovation may leap over every limit, but this raises new dangers rooted in our technological prowess. Technological advance threatens us in three ways: by causing political instability, as society is unable to successfully adapt to new technologies; through the rise of totalitarianism, as advances place the means for surveilling, manipulating, and controlling the population in the hands of governments; and by causing social or environmental disruption, via either a catastrophic accident or the intentional actions of rogue actors (Frodeman, 2019).

Transhumanism highlights the fact that the overall results of knowledge production take us in a direction quite different from the piecemeal outcomes of these efforts. Those who dismiss transhumanism do so by focusing on the piecemeal aspects of our culture of knowledge. Heidegger called this the forgetfulness of Being—the loss of a sense of our overall trajectory as we focus on smaller matters. Across the modern era regional ontology has trumped fundamental ontology as small questions have stood in the place of large ones. Amusements pile up even as civic virtues fade. With all our riches we have created a trivial culture.

## 4.7 Sustainability and the University

Great questions assert themselves in the environmental crises we face. As I have noted, many of these crises are rooted in scarcity—pollution (including $CO_2$) being a matter of not enough land or water or air to disperse contaminants, and extinction resulting from not enough space to support wild species. As the science of limit, helping us to understand where planetary boundaries lie, geology should be the sovereign of our epistemic empire.

The account offered here has connected geologic knowledge and perspectives to a fundamental rethinking of the premises underlying our culture of knowledge. The points made are speculative in nature. The scenario described—where a more philosophical and policy-oriented field of geology becomes the culmination of our educational efforts, as well as the governor of our research efforts—is not yet plausible. But one of the roles of philosophy and of intellectual work generally is to sketch out possible futures, knowing that most of these futures will not come to pass. Such efforts can still be worthwhile. The function of a thought experiment is sometimes to help forestall a future by sketching out its undesirable dimensions. And these efforts may not be entirely utopian: the recent effort in Chile to rewrite the country's constitution from an ecological standpoint—giving nature rights and considering the needs of future generations—is a sign that change is possible (New York Times, 2021).

In a previous work (Frodeman, 2019) I argued that the most likely driver of the shift in intellectual culture called for here would be a medium-sized societal catastrophe. An event where perhaps 5% of the world's population died through a disaster rooted in either environmental crisis or technoscience run amok might prompt the rethinking of our epistemic assumptions. No one desires such a scenario. But Nietzsche may be correct that people acquire new mental habits only through an event painful enough to etch it in their memory. In the meantime, intellectuals make arguments and artists create works in the hope that they may launch a movement or persuade people in positions of power.

As it happened, those speculations have been mirrored by subsequent events. The last two years (this is being written in early 2022) has seen both significant environmental disruption via weather events and the rise of a pandemic that may have resulted from gain of function research that escaped the laboratory. We do not know if the recent bizarre weather is merely the start of massive changes in the climate. Nor do we know if COVID-19 originated in the Wuhan Institute of Virology or have a clear grasp of the societal changes that will occur in the wake of COVID-19. But the early signs are that society seems to have become more dysfunctional rather than using these crises to re-evaluate its behavior.

Even if the suggestion of a new geology-based epistemology sounds far-fetched, it is clear that we are facing epistemic disruption of one kind or another. The function of the modern research university, where it creates, certifies, and disseminates knowledge, is under siege. For 150 years the university has been the uniquely authoritative source for knowledge. Today, however, the university's central role in knowledge culture has been undercut by the rise of the Internet. Web 2.0 and social media have created alternative epistemic spaces that have undermined the role of expertise. This has contributed to a wide range of results, including vaccine skepticism and the rise of rightwing authoritarian political movements in a number of countries.

By way of conclusion, let us note some of the possible consequences of the perspectives offered here. In terms of the university, the current grab-bag, horizontal structure could give way to a hierarchical focus structured in terms of sustainability. Environmental change courses could become part of our intellectual grammar and frame the overall goals of intellectual work. The brightest high school students would take advanced placement courses in geology rather than in physics and calculus. Such courses (for there would need to be more than one) would be complemented by geoethics courses that would be inter- and transdisciplinary in nature, moving from science to risk assessment to restorative justice and back again. The main point of all these efforts would be to recognize that we now have a common end that should transcend all our other values: the protection of our planet.

Concerning our research portfolio, this argument implies moving beyond the libertarian epistemology that has underlain the academy for the last 150 years. Those epistemic pursuits that support a sustainable way of life would be pursued. Those epistemic efforts that will exacerbate our current unsustainable trajectory would be restricted, banned, or go unfunded.

This does not imply an epistemic authoritarianism. Reorienting university life and society generally toward the goal of sustainability should be a matter of persuasion and nudges more than regulation, prompting the slow process of changing the *Zeitgeist* of a culture. People will disagree about the nature of a given project and will argue whether the attached harms are trivial or are offset by positive results. There will be debates and compromises; people will disagree on interpretations. All of this is appropriate within democratic societies. The point is one of framing: there would be a general recognition that protecting the environment and observing its limits is the paramount public good of all our activities.

These points have been put in terms of geology partly in recognition of the fact that we live in a scientific era. But ultimately the change in worldview being called for here is psychological, philosophical, and spiritual in nature. The long history of humanity has been shaped by want. Chronic lack has molded our psyches to always want more. This has reached such absurd heights that men with hundreds of billions of dollars still seek to augment their wealth. It is time to leave the adolescence of humanity behind and create a culture of maturity.[1]

# References

Clark, W. (2006). *Academic charisma and the origins of the research university*. University of Chicago Press.

Frodeman, R. (2014). *Sustainable knowledge: a theory of interdisciplinarity*. Palgrave McMillan.

Frodeman, R. (2019). *Transhumanism, nature, and the ends of science*. Routledge.

Johnson, K., Vriesen, J., Stabb, G., & Braginetz, D. (2005). *Ancient Denvers: Scenes from the past 300 million years of the Colorado Front Range*. Chicago Review Press.

Laudan, R. (1987). *From minerology to geology*. University of Chicago Press.

Lyotard, J. -F. (1984). *The postmodern condition: A report on knowledge*. University of Minnesota Press.

McPhee, J. (1983). *In suspect Terrain*. Farrar, Straus, Giroux.

New York Times. (2021). 'Chile rewrites its constitution, confronting climate change head on', Somini Sengupta, December 28, 2021, at https://www.nytimes.com/2021/12/28/climate/chile-constitution-climate-change.html. Accessed September 21, 2022.

Peppoloni, S., & Di Capua, G. (2015). The meaning of geoethics. In S. Peppoloni, & M. Wyss (Eds.), *Geoethics: Ethical challenges and case studies in Earth Science* (pp. 3–14). Elsevier. https://doi.org/10.1016/B978-0-12-799935-7.00001-0

Shattuck, R. (1997). *Forbidden knowledge: From Prometheus to Pornography*. Mariner Books.

Shepard, P. (1982). *Nature and madness*. University of Georgia Press.

---

[1] Other versions of some of the points made here have appeared in earlier publications (e.g., Frodeman, 2019).

# Chapter 5
# The Ethics of Gaia: Geoethics From an Evolutionary Perspective

Sofia Belardinelli and Telmo Pievani

*Man has been here 32,000 years. That it took a hundred million years to prepare the world for him is proof that that is what it was done for. I suppose it is. I dunno. If the Eiffel tower were now representing the world's age, the skin of paint on the pinnacle-knob at its summit would represent man's share of that age; and anybody would perceive that that skin was what the tower was built for. I reckon they would, I dunno.*
*Mark Twain*

**Abstract** In times of unprecedented ecological change led by human activities, a global ethical framework is most needed to support the rapid transformation of current development models, to ensure the protection of human and non-human nature. Geoethics offers such a universal system of values. We assess to what extent geoethics maintains an anthropocentric perspective and examine the ethical challenges raised by this statement, arguing that (i) geoscientific knowledge, which investigates the interrelations between the biotic and abiotic world in a deep-time perspective, should imply the adoption of an eco-centric or even geocentric perspective; (ii) the assumption of an anthropocentric perspective should be outlined more precisely, by clarifying the utilitarian and deontological reasons to maintain a weak anthropocentric approach and to avoid the theoretical bias underlying many anthropocentric narratives. Then, we claim that a non-anthropocentric geoethics would allow a better understanding of the role *Homo sapiens* as a species plays within the biosphere and geosphere. To provide evidence for our hypothesis, we will discuss two case studies: (i) climate change as a monumental niche construction process, where the ambivalence of human nature is seen in action; (ii) the co-evolution and interconnection between biological and cultural diversity, which support each other in an inextricable link. We claim that a humanistic, eco-centric geoethics can support

S. Belardinelli (✉)
Department of Humanities, University of Naples Federico II, Naples, Italy
e-mail: sofia.belardinelli@unina.it

T. Pievani
Department of Biology, University of Padua, Padua, Italy
e-mail: dietelmo.pievani@unipd.it

© The Author(s), under exclusive license to Springer Nature Switzerland AG 2023
G. Di Capua and L. Oosterbeek (eds.), *Bridges to Global Ethics*,
SpringerBriefs in Geoethics,
https://doi.org/10.1007/978-3-031-22223-8_5

the necessary transition towards a new conceptual framework in which humans are not separate from, but part of the biosphere. This approach is part of the philosophy of biology, a field that programmatically converges humanities and the life sciences.

**Keywords** Environmental ethics · Anthropocentrism · Biocultural diversity · Climate change · Responsibility

## 5.1 Introduction

The multiple crises Earth and its inhabitants are currently facing are mainly human induced, and the rapid worsening of the environmental crisis gives us less and less time to act. Coupled with the adoption of urgent measures of adaptation and mitigation, a wise application of available knowledge and technologies is an important part of the solution in the short term.

Geoethics has been developed to address these challenges, yet it is defined as an "inevitably anthropocentric ethics" (Peppoloni & Di Capua, 2021c). Although avoiding an *anthropogenic* approach can be difficult when considering the human-nature relationship during the Anthropocene, this does not necessarily imply that a strong *anthropocentric* perspective should be accepted. Indeed, there is a fundamental difference between a minimal degree of anthropocentrism (which is unavoidable due to cognitive and evolutionary reasons) and the anthropocentric prejudice that justifies the belief in a supposed human superiority. We address the ethical challenges raised by the adoption of an anthropocentric perspective, albeit in a "weak" understanding, through an interdisciplinary lens that understand the life sciences from a philosophical perspective. Then, we give strength to our statements by discussing some case studies.

From an evolutionary perspective, we focus on the ambivalent nature that characterises our species, which is at the same time a geological force of change and a "hyper-keystone species," fundamental to the well-being of different ecosystems. Through the lens of evolutionary biology, we describe *Homo sapiens* as a niche-constructing species, acting on a planetary scale. This attitude can lead to environmental collapse or to a pacific cooperation with other inhabitants of Earth.

We claim that a humanistic, eco-centric geoethics can support the necessary transition towards a new conceptual framework in which humans are not separate from, but part of the biosphere. Therefore, responsibility towards the Earth does not derive from a supposed ontological superiority, but from our incidental power to destroy or preserve nature in its existing state. This approach is part of the philosophy of biology, a field that programmatically converges humanities and life sciences.

## 5.2 Critique of Anthropocentric Geoethics

Geoethics has its roots in the field of geosciences. Originally developed as a professional deontology (Peppoloni & Di Capua 2021a), it is defined by the IAPG as «the research and reflection on the values which underpin appropriate behaviours and practices, wherever human activities interact with the Earth system» (Peppoloni & Di Capua 2015; Di Capua et al., 2017; Peppoloni et al., 2019). Initially intended only for geoscientists, over time, geoethics has broadened its focus and outreach. Today, this new discipline aims not only at providing geoscientists with a practical code of conduct to face ethical dilemmas, but also attempts to build a global framework of universal values for the Anthropocene—"the human epoch" (Crutzen & Stoermer, 2000; Ellis, 2018a).

Humans have profoundly modified their environment for at least 12,000 years (Ellis et al., 2021), but in recent times, the impact of human activities on the biosphere has become significantly heavier. As shown by Elhacham and colleagues (2020), in 2020, the mass of human artefacts has exceeded all living biomass. We humans, accounting for a mere 0.01% of the global biomass and weighing less than all the existing bacteria, have crowded the world with 1.1 teratonnes of artefacts. This is the overwhelming footprint of the Anthropocene on Earth. Faced with this unprecedented situation, accelerated during the last century (McNeill & Engelke, 2014), new ethical foundations are needed to steer the unequal relationship between humans and the rest of the natural world towards a just transition.

Besides values such as honesty, respect and justice, the core value of geoethics is responsibility (Peppoloni & Di Capua 2022b, pp. 31–47). Since human actions have such a serious impact on the biosphere and geosphere, we are called upon to respond to those consequences. In its relationship with the Earth system, the human species has acquired enormous power, and this implies great responsibility in preserving the biotic and abiotic elements of the planetary system. Geoethics is essentially «actor-centric, and in particular oriented towards informing the conceptual frameworks and practical interventions of the individual scientist» (Peppoloni et al., 2019). The individual, through her or his specific knowledge, has the duty to promote «cooperation with those who are not experts in the field, to find the most acceptable ways in which to interact with the Earth system» (Peppoloni et al., 2019).

This global ethical framework is therefore addressed not only to those directly involved in geosciences, but to every member of human society, who are called upon to recognise their responsibility towards every human being and non-human nature.

One of the key theoretical foundations of geoethics lies in the identification of the recipient of this ethical proposal: according to Peppoloni and Di Capua (2021a), the fact that humankind is a geological force requires an ethical framework centred on human agency and oriented towards a responsible relationship with the Earth system. According to the authors, such an ethical proposal «inevitably leads to an anthropocentric perspective» (Peppoloni & Di Capua, 2021a).

It is an undeniable fact that the current global climatic and environmental crises have an anthropic origin. Scientific evidence shows that human actions are comparable to a geological force, able to modify and shape the Earth system. Geoethics emphasises the ethical duty to recognise—as individual and collective human agents—this exceptional role and to take action responsibly.

According to Peppoloni et al. (2019), «The unavoidable reality of anthropogenic change makes a degree of anthropocentrism *a necessity*» (our italic). Though substantially agreeing with this thesis, we suggest that the notion of anthropocentrism should be better explained, to mark the difference between a minimum degree of unavoidable cognitive anthropocentrism, which is related to the way we humans experience and describe to ourselves the outside world (Peppoloni & Di Capua, 2021b) and a strong anthropocentric perspective, which is the only one bearing ethical consequences, and thus should be carefully avoided. From this viewpoint, an anthropocentric ethical proposal is always unacceptable, as its defence is slippery and might lead to the unwanted recognition of qualitative differences between the "actor" and the recipients of action. Thus, our critique of a—even weak—anthropocentric geoethics consists of two main issues, which will be discussed in the following sections:

(i)   geoscientific knowledge, which investigates the interrelations between the biotic and abiotic world in a deep-time perspective, might take advantage from the adoption of an eco-centric or even geocentric perspective, albeit with a minimum (inevitable) degree of weak, cognitive anthropocentrism;
(ii)  the supposed unavoidability of an anthropocentric point of view could be the result of a theoretical bias and has to be carefully defined.

### 5.2.1  Geosciences are Non-anthropocentric by Nature

The geosciences are a diverse set of disciplines that study different aspects of the Earth systems, from their formation to their internal functioning and dynamics.

Over time, geoscientific knowledge has contributed to change and extend our ordinary perception of Earth dynamics, showing the complexity and interconnection of the different planetary realms—physical, chemical, biological and geological, as well as socio-economic and cultural. Scientific concepts such as deep time, physical laws and evolutionary processes offer the possibility of radically changing the way humans perceive the world. These notions call attention to the fact that, within the complex biogeochemical systems of the planet, the young mammalian species *Homo sapiens* does not play a pivotal role per se. Evolutionary biology, among other disciplines, has conclusively shown that human beings are not separate from nature, but rather inherently dependent on it. Moreover, the clash between human time and geological deep time irreversibly changes the perception of our place in history, unveiling our fragility in the history of the Earth (Pievani, 2015).

Thus, a science-based understanding of the Earth and its internal dynamics implies a shift away from a strict anthropocentric point of view: this is, in fact, inconsistent with scientific evidence, which demonstrates on the one hand the marginal role of our species in the complex planetary systems, and on the other hand, the need to assign equal importance to all the biotic and abiotic components that make up the biosphere and geosphere. Following the late palaeontologist Stephen J. Gould (1941–2002), we acknowledge that the existence of our species was not a necessary outcome of the history of life, and that it is not fundamental to the survival and well-being of the biosphere; on the contrary, we depend on it for our very existence. SARS-CoV-2 pandemic, and our co-evolution with the variants of this new virus, is a dramatic evolutionary confirmation of our vulnerability. Geology and evolutionary biology contributed to unveil this shocking truth: «By the turn of the last century, we knew that the Earth had endured for millions of years, and that human existence occupied but the last geological millimicrosecond of this history. […] If humanity arose just yesterday as a small twig on one branch of a flourishing tree, then life may not, in any genuine sense, exist for us or because of us. Perhaps, we are only an afterthought, a kind of cosmic accident, just one bauble on the Christmas tree of evolution» (Gould, 1989, p. 44). We suggest that this scientifically informed perspective, based on the complexity of evolutionary history from a deep-time geological point of view, is more in line with an eco-centric or even geocentric position, rather than with anthropocentrism.

In harmony with this vision, geoethics acknowledges every element of the Earth's ecological and geological systems as ontologically equal, postulating «the duty to guarantee to any entity the same value and existential space based on a recognised diversity» (Peppoloni & Di Capua, 2022b). Traditional anthropocentric views attribute to non-human nature a mere instrumental value, defined by its usefulness for human socio-economic systems. Geoethics, by contrast, acknowledges the essential ontological value of all biotic and abiotic elements of life, thus intuitively overcoming the anthropocentric claims.

Although not ontologically different from any other, our species does have a unique feature, that is the ability to understand the functioning of the external world and to grasp its own existential condition. Thus, the attempt to detach from our natural anthropocentric perspective and to adopt an eco-centric point of view is not, again, an arrogant claim of superiority or the symptom of irrepressible misanthropy. Instead, this balanced eco-centric perspective is supported by evolutionary knowledge, which has provided us with the means to understand that we do not play a pivotal role in the biosphere.

In a sense, these two dimensions, the eco-centric and the anthropocentric, are not mutually exclusive (Peppoloni & Di Capua, 2022a). In fact, our commitment to revert the current trend of biological and climatic collapse is, first and foremost, an autonomous human decision aimed at ensuring our own survival and that of other species, knowing, however, that this will not be a conclusive action since life would still thrive in other forms after the extinction of *Homo sapiens* (Pievani & Lanting, 2019, p. 32).

### 5.2.2 Examining the Anthropocentric Claim: A Theoretical Bias?

The alleged inevitability of an anthropocentric perspective for geoethics does not seem to have sufficient theoretical justification. Rather, considering anthropocentrism as "the only possible alternative" could be the result of a theoretical fallacy, which consists in the overlap of the notion of anthropo-*centric* on that of anthropo-*genic*.

It is clear, in fact, that we, as human beings, will never be able to look at the world around us in (albeit minimal) non-anthropocentric terms since our knowledge intrinsically depends on human experience and cognitive filters. However, this does not imply the a priori inability to assign an objective value to non-human nature and to recognise ontological equality to all living and non-living components of the Earth systems, from an axiological perspective. We thus hypothesise that the anthropocentric claims of geoethical discourse might be affected by cognitive biases and therefore need to be carefully examined.

Within the field of environmental ethics, anthropocentrism has been interpreted in different ways. On the one hand, there is a "weak" and pragmatic understanding of the concept, which emphasises the fact that our actions aim (or should aim) primarily at ensuring the survival of our species but, at the same time, recognises the importance of not damaging the delicate balance of the ecosystems we are part of. On the other hand, there are "strong" anthropocentric perspectives that are subject to the cognitive bias which portrays our species as more relevant than the others in the "economy of nature". This applies, for instance, to the cosmological "anthropic principle" in its strong version, according to which the current well-ordered composition of the universe, which has been crucial for the emergence of life, could not have originated by chance. This argument, however, is a truism, as it does not demonstrate that it was necessary for the history of the universe to unfold as it did. As Stephen Hawking (1942–2018) stated in one of his well-known popular science books, «the strong anthropic principle would claim that this whole vast construction [the Universe] exists simply for our sake. This is very hard to believe. Our Solar System is certainly a prerequisite for our existence, and one might extend this to the whole of our galaxy to allow for an earlier generation of stars that created the heavier elements. But, there does not seem to be any need for all those other galaxies, nor for the universe to be so uniform and similar in every direction on the large scale» (Hawking, 1988, p. 130). Similarly, when it comes to anthropocentrism in the strong sense, our inevitably biased and partial comprehension of reality should not lead to ethical generalisations.

A third element that is relevant to grasp the difference between a strong and a weak anthropocentric perspective is the concept of *anthropogenic*. Human capacity to understand their place in the Universe implies the ability to recognise the way in which our activities have an impact on the rest of the biosphere. This ability to acknowledge the anthropo*genic* nature of the major changes that have recently occurred in the dynamic ecological equilibria of the biosphere demands the prompt assumption of responsibility towards our descendants and other living beings.

We analyse whether, as is in the case of the "anthropic principle," the geoethical anthropocentric standpoint, albeit adopting a revised and weakened anthropocentrism, may still be affected by the "anthropocentrism bias" that underlies modern Western philosophy (Naudé, 2017). To challenge the dogmatic Western conception of the human as the centre of all ethical and theoretical reflections concerning nature and the world is a necessary step to overcome the traditional strong anthropocentric perspective. The first action we can take to achieve this aim is, according to Naudé (2017), «the decentring—in our minds and consequently in our actions—of the human subject as superior apex- and reference point of the natural order. We stand in and with the natural order, as intrinsically part of, and not above, this order».

A crucial role in undermining strongly anthropocentric claims is played by evolutionary sciences. In fact, recent theories and evidence coming from this field help redefine the traditional teleological understanding of the history of life, e.g. by replacing the view of evolution as a steady and linear progress towards greater complexity and perfection with a humbler and more realistic "tree of life," an intricate web of branches that develop simultaneously, following specific natural laws without a predetermined purpose. This is also true for the complex phylogeny of hominin species.

Geoethical thinking is strongly consistent with this evolutionary view, as it does not embrace an ethical anthropocentric position. In fact, it does not place the human being at the centre of reality, and moreover, it confers on humans not rights but duties towards the Earth. As Peppoloni et al. (2019) state, «The *Anthropos* is assigned the unconditional responsibility of being part of a whole and an equal among all. This perspective assigns to the human being a centrality in the Earth System in terms of responsibility and not of dominance and power». We argue that for humanity to recover from the current crisis an ethical proposal is required that is centred on human as an active agent, not as a recipient of rights.

Geoethics has been characterised as a virtue ethics (Peppoloni et al., 2019): it proposes a set of virtues and shared values on which to base our actions to protect and preserve non-human nature, to which our respect and responsibility must be oriented. Adopting the weak anthropocentric perspective, which recognises the minimal anthropocentric nature of our ethical reflections but does not assign differential ontological value to human beings and other inhabitants of Earth, is the first (necessary, though not sufficient) step towards the foundation of a new ethical framework. Its core values are humility, rooted in a science-informed awareness of our marginal role in the history of life, and responsibility, which translates into unconditional duties to future generations and to life as we know it.

Parenthetically, it is important to clarify that the eco-centric ethical viewpoint here proposed should not be conflated with a misanthropic perspective. There are four main reasons to refuse an anti-humanist narrative, which usually depicts the human species as a dangerous "virus" or "cancer" for the Earth. These reasons are rooted in the ecological and evolutionary-based awareness of our marginality within the planetary system; nonetheless, we are still a biological species and therefore subject to evolutionary rules.

(i) First, from this point of view, the natural disposition to satisfy our basic needs and to protect the "evolutionary interests" of our species need not be labelled as a strongly anthropocentric attitude since any species tends to maximise its evolutionary fitness;

(ii) Second, the geoethical virtue of responsibility must be directed primarily to future human generations, whose fundamental right to live healthy should be protected from the likely negative outcomes of the current ecological crisis;

(iii) Moreover, we have to acknowledge the complex historical responsibilities involved in the emergence of the climatic and environmental transition. These events result from the activities carried out by a specific human society and not by the species as a whole; a misanthropic attitude does not take into account the demands for distributive justice coming from the so-called "Global South;"

(iv) Lastly, the inherent ambivalence of human experience and creativity carries a complex and multifaceted legacy, made not only of environmental disruption but also of a wealth of artistic and cultural creations, which may be worth preserving.

## 5.3   Humans' Place in the Earth System: A Non-anthropocentric Ambivalence

A perspective that refuses strong anthropocentrism rejects any claim of ontological disparity or superiority of the human species over the rest of the living world. However, empirical evidence suggests quite the opposite: human actions have now acquired an unprecedented influence over the natural world, and how we manage this power will determine the fate of the natural realm as we know it.

Our species is marked by a profound, yet unresolvable, ambivalence. On the one hand, *Homo sapiens* is a young species whose evolutionary history has been shaped by natural events and contingencies, and whose complex socio-economic systems are still today, despite scientific and technological advances, totally dependent on natural resources. Furthermore, *Homo sapiens* is a crucial element for the ecosystems it has shaped, and to which has adapted over time—it can be defined as a "hyper-keystone species" (Worm & Paine, 2016; Sullivan et al., 2017). On the other hand, however, the peculiar evolutionary trajectory of *Homo sapiens* has led to the predominance of a single species over almost every ecosystem on Earth, probably an unprecedented event in the history of life.

The animal *Homo sapiens* is, as every other living being, inherently bound to its environment, on a microscopic as well as on a macroscopic scale. The concept of the *holobiont* clearly describes this perspective from a molecular point of view (Gilbert & Epel, 2015): every individual is a member of a broader ecosystem and, at the same time, an entire ecosystem itself, hosting a wide variety of living organisms. Moreover, humans depend on nature also on a wider, societal level. Indeed, today's complex human societies could not exist if not relying on fundamental "ecosystem services," such as insects' pollination, the "carbon sink" role of forests and oceans, the cycle of nutrients which fertilizes the land and many others (Dasgupta, 2021).

Nonetheless, human expansion is now resulting in increasingly high environmental costs. Despite its role as a keystone species, anthropic interventions have often been destructive for planetary ecosystems, recently leading to the current environmental crisis. The extreme niche construction processes carried out by humans profoundly altered the complex and dynamic equilibria of Earth's systems and cycles, ultimately triggering a rapid and erratic change in the stable circumstances on which our own species has relied for the last millennia. This has led to an unprecedented situation: *Homo sapiens* could be provocatively considered as the first "self-endangered species" (Pievani & Meneganzin, 2020).

Such complexity has prompted conflicting interpretations of the human-nature relationship. Some, emphasising its destructive potential, underline the differences that separate humans from the rest of the natural world; others, on the contrary, highlight the long evolutionary interdependence between *Homo sapiens* and other life forms, thus supporting the "naturality" of our species. In a sense, both these explanations are correct, and human nature cannot be understood without considering this ambivalence, which is also crucial to understand our complex evolutionary relationship with the environment. As stated by Sullivan et al. (2017), embeddedness and destructiveness coexist in our species: «At local levels, co-evolutionary relationships between small-scale human societies and non-human taxa in the same community can be so longstanding and immersive that the removal of humans results in simplified food webs or ecosystem collapse, while in the context of global population growth the intensity of human–environment interactions has led to numerous documented cases of wildlife population decline and extinction».

### 5.3.1  Destructive or Fundamental to the Ecosystems? Two Case Studies

In order to assess this complex issue, we discuss two case studies, each providing evidence in support of one of the previously presented theses. The first case study is the current climate breakdown: the rapid systemic transitions taking place on our planet are indeed the outcome of a globally extensive process of niche construction consisting of prolonged activities for the adaptation to human needs, which resulted in heavy environmental alteration.

Drawing on a historical reconstruction of human-nature interactions, the second case study analyses the interdependence of biological and cultural diversity, two domains which share a long history of co-evolution. Indeed, the manifestations of biocultural diversity disseminated across the planet show that, since their emergence as a species, humans have been not only a disruptive invasive species, but also, on the long term, a key driver for the preservation and multiplication of biodiversity.

In both cases, it is evident that human-nature relationships cannot be understood without taking a deep-history perspective into account. The history of co-evolution between *Homo sapiens* and its multiple environments carries a complex and ambivalent legacy both from an evolutionary and a cultural point of view.

On the one hand, throughout human history, our long-term extreme niche construction activities have had profound impacts on newly colonised ecosystems around the world, and have recently resulted in a global climate transition that is now affecting ecological systems worldwide, posing a threat even to the well-being and survival of our own species. On the other hand, however, in many cases, humans have been—and continue to be—a fundamental element for the ecosystems they live in, sustaining the existence of hundreds of endangered species with their traditional cultural knowledge. Present-day biodiversity, despite drastically diminished and reshaped by the colonisation of human prehistoric populations in a remote past, is to a large extent the result of millenarian processes of co-evolution with human populations, distributed across every environment on the planet. Today, however, this mutual balance is under threat.

### 5.3.1.1 Climate Change: The Unintended Outcome of Human Niche Construction Strategies

Following Meneganzin et al. (2020), we interpret the anthropogenic climate crisis as the result of a "monumental niche construction process" undertaken by humans. The human endeavour to manipulate the environment to meet the cultural needs of our species can be understood as an evolutionary process, as it has profound consequences on the world's ecosystems and affects the evolutionary paths of many species.

Niche construction is an evolutionary strategy through which «the organism influences its own evolution, by being both the object of natural selection and the creator of the conditions of that selection» (Levins & Lewontin, 1985). It is now undisputed that this biological process has played a central role in human evolution. Indeed, throughout their evolutionary history, humans have improved this strategy by adding a fundamental cultural component to it. As stated by Laland et al. (2016), «Niche construction due to human cultural processes can be as potent as niche construction that has evolved through biological evolution». For our species, the cultural dimension is a primary driver of evolution: by modifying the environment to suit our socio-cultural demands, rather than our biological necessities, we initiate substantial and long-lasting modifications and pass on to future generations not only new knowledge and skills, but also a novel social context that, in turn, represents a constraint for the subsequent evolutionary development of the species.

Human niche construction activities are grounded in a vast amount of cultural knowledge, which allows us to modify the pre-existing environmental selective pressures at a pace and scale previously unknown for any other living species. This cultural inheritance includes both non-material and material aspects (Adger et al., 2013), consisting not only of the knowledge and information accumulated over time by different societies and cultures, but also of constantly ameliorated innovations and technological skills, which play a central role in the way humans interact with their environment. This broad cultural base, coupled with a blatant short-sightedness in decision-making, has recently led to unparalleled consequences, such as unintended global climate change.

Our niche construction strategy is not substantially different from that of many other species, which rely too on cultural adaptation. Our actions, however, develop in a wide variety of different ways, thus influencing the evolution of our own species, as well as shaping ecological environments «by imposing artificial selection, or through direct genetic modification, to fashion the course of biological evolution for numerous other species. In these respects, human cultural niche construction is unique» (Odling-Smee & Laland, 2011). Our adaptation efforts aim not only at promoting individual fitness and ensuring our survival as a species, but also at gaining a more comfortable life and satisfying our cultural desires. In striving to achieve these goals, the human niche has changed rapidly over time, thus requiring more and more innovations and adaptations. The process is unceasing because «the inherent goal is not to reach an equilibrium with the environment to sustain fitness, but rather to innovate for other purposes» (Low et al., 2019). Therefore, as low and colleagues argue, humans are to be considered niche *modifiers* rather than niche constructors, as their environmental alteration activities are relentless and do not only pursue adaptation, but aim at an endless series of cultural and technological innovations (Low et al., 2019).

However, this pervasive niche construction (or modification) process is having unintended consequences today. We have altered our ecological niche to such an extent that we are becoming less adapted to it (Xu et al., 2020). Moreover, our cultural and technological innovations are producing a "negative" ecological inheritance, that is, we are passing on to our descendants an ecological, social and cultural environment that is far less hospitable than the one we are living in. In other terms, our culturally driven actions have turned out to be maladaptive, as anthropogenic climate change now threatens the very survival of our species and is likely to leave behind poorer and far more unequal human societies. In turn, the adoption of mitigation and adaptation strategies might have profound impacts on the cultural inheritance of communities and societies worldwide, possibly leading to reduced resilience and, in general, to maladaptive outcomes.

As stated by Meneganzin et al. (2020), the world is now witnessing a pivotal moment, in which the choices we make will be crucial in shaping the near and remote future for ourselves and many other living species: «The one that is unfolding before our eyes is a high-risk evolutionary experiment, in which one species will have no choice but to produce an adaptive response to the environmental changes that have been triggered by its own activities. The available window for action to mitigate climate change, in order to prevent the deleterious impacts from growing and involving an increasing number of generations after ours, is closing rapidly». Since our actions have created the conditions for the alteration of our ecological inheritance on the long term, an ethical commitment is now imperative. Indeed, the physical, ecological, social and cultural consequences of human activities cannot be ignored, and transformative action is now required as a moral duty, based on the non-anthropocentric recognition of the urgency to guarantee every living being the right to live.

### 5.3.1.2   Biocultural Diversity: Coexistence is Possible

Niche construction—or niche modification, if we are to follow some scholars—is not in itself destructive. In fact, the human ability to actively interact with the environment by transforming it has been an important driver of evolutionary innovation. In other terms, throughout their evolutionary history, humans have contributed to increase global diversity. The extent of this contribution can be easily assessed by analysing biocultural diversity, which is «the sum of the world's differences, no matter what their origin. It includes biological diversity at all its levels, from genes to populations to species to ecosystems; cultural diversity in all its manifestations (including linguistic diversity), ranging from individual ideas to entire cultures and, importantly, the interactions among all of these» (Loh & Harmon, 2005).

A growing number of studies highlights the profound interdependence between biodiversity and the richness of human cultural manifestations. Even the two extinction crises currently taking place worldwide—namely, the global reduction of biodiversity and that of languages and traditional cultures—prove the existence of a deep connection between these two domains. Moreover, field surveys, global and regional maps and quantitative assessments document the globally widespread overlapping of biological, cultural and linguistic diversity hotspots. Therefore, those different forms of diversity may have had a common origin and might as well rely on one another for their development and preservation (Gorenflo et al., 2012).

Interestingly, many places commonly recognised as "sanctuaries" of biodiversity share a long history of stable coexistence with human populations, which are often stewards of this natural heritage. Indigenous peoples and traditional cultures retain an inestimable wealth of "traditional ecological knowledge" (TEK), which is the result of millennia of coexistence and co-evolution with other components of a specific ecological niche (Maffi, 2018).

In many cases, human activities have been crucial not only for the preservation but also for the increase of local biological diversity. A striking example of the importance of human activities for the ecosystems comes from Australia: there, the practice of "cultural burning," vital to Australian Aboriginal peoples, has been recognised as fundamental, still today, in shaping and preserving the desertic ecosystem and its inhabitants. Although, as several studies suggest, in ancient times human colonisation of the continent likely triggered the extinction of the local megafauna, in this case there is «strong evidence for significant ecological adaptation to long-term keystone presence of intensive human disturbance, which could include adaptive morphological evolution in non-human taxa», as Sullivan and colleagues (2017) stress out.

Another example that well illustrates the interconnection between human and non-human nature comes from Italy: in the last 70 years, the country's inland mountain areas, whose ecosystems have been shaped by human agricultural activities for millennia, rapidly depopulated. As the rural population decreased, biological diversity declined as well. Indeed, *Homo sapiens* was a keystone species for those environments, and its absence led to a sudden disruption of the region's ecological network (Fulgione & Troiano, 2019; Troiano et al., 2021). Thus, human presence can be either a driver of environmental stress—as the climate crisis demonstrates—or a core resource for the conservation of ecosystem.

In parallel, in recent years, scholars have witnessed that wherever a biodiversity hotspot declines, indigenous peoples suffer, and native languages, traditions and knowledge are increasingly lost. This trend has been observed in biocultural diversity hotspots all over the world, which are paying a heavy price for the consequences of deforestation, defaunation (Dirzo et al., 2014), ecosystem disruption and climate change.

This vast array of data clearly indicates the existence of an "inextricable link" between human and non-human nature: our species cannot legitimately be regarded as ontologically different from the rest of the natural world. Moreover, the evolutionary history of *Homo sapiens* proves its ability to coexist with other living beings within the biosphere. The current maladaptive outcome of our actions is not proof of our separateness from nature; instead, it confirms the destructiveness of the exploitative socio-economic model that has become predominant in recent times.

The biocultural co-evolution of peoples and ecosystems dates back to at least 12,000 years ago (Ellis et al., 2021), and throughout this period of time, the world's biodiversity, though modified and shaped by human activities, has continuously thrived. «Although some societies practicing low-intensity land use contributed to extinctions in the past, the cultural shaping and use of ecosystems and landscapes is not, in itself, the primary cause of the current extinction crisis and neither is the conversion of untouched wildlands, which were nearly as rare 10,000 years ago as they are today. The primary cause of declining biodiversity, at least in recent times, is the appropriation, colonisation, and intensifying use of lands already inhabited, used and reshaped by current and prior societies» (Ellis ct al., 2021).

### 5.3.2 The Unavoidable Ambivalence of Human Nature

The two case studies discussed above are complementary to each other and help us outline some key features concerning the inherent ambivalence of human-nature relationships.

First, the long history of biocultural co-evolution and the current outcomes of human niche construction processes suggest that, from an ethical perspective, a sought-after "return to nature" is neither possible nor desirable because it would entail the existence of an ontological separateness between humans and the rest of nature. Instead, natural sciences show that humans are embedded *in* the natural world.

This is also proved by the fact that a "pristine" nature—that is, a natural environment free of any human traces—has barely existed on the planet for many millennia now. Hence, a misanthropic ecocentrism, such as the movements supporting the extinction of our species "for the sake of the planet", lacks any scientific basis, as it ignores the complex interrelationships that inseparably tie the human species to the rest of the biosphere.

In addition, both case studies highlight the radical ambivalence of the role humans play in the ecosystem they inhabit. Throughout its evolutionary history, *Homo sapiens* has been both destructive (as it is at the global level today) and fundamental (as is still the case for many indigenous peoples and traditional communities). The ability to drive our own evolutionary path by cultural means binds us to long-term adaptation processes which, in turn, might prove maladaptive in the face of a rapidly changing environment.

Therefore, the assumption of an ethical perspective which has its core in the value of responsibility is required. This geoethical virtue—as Peppoloni and Di Capua (2021a) name it—needs to be directed towards future human generations as well as other living beings. The main ethical purpose for the Anthropocene is thus to guarantee the perpetuation of life as we know it.

As suggested above, geoethics does not need to completely abandon an anthropocentric perspective. In fact, this is not possible since we humans can only look at reality from our own partial point of view. We thus propose that geoethical thinking, which aims to offer universal ethical guidance, adopts a weak and "disenchanted anthropocentrism." This means the following:

(i) Acknowledging that our understanding of reality is inevitably biased by our cognitive and evolutionary inheritance, and accepting that we will always try to meet our own needs first;

(ii) Recognising, informed by scientific evidence coming from evolutionary biology, ecology and other disciplines, that we are not necessary in the "economy of nature," and that the biosphere would still thrive after the extinction of humans;

(iii) Drawing from this evidence, taking responsibility to ensure that human actions do not violate the right to live of future generations and do not adversely affect the survival and well-being of other living species, which are accorded ontological equality.

## 5.4 Conclusions

As shown by the ongoing pandemic, the lack of responsibility in managing our technologies and power towards nature has proved detrimental even to human societies. Along with the climate crisis, we are entering a "pandemic era" (Morens & Fauci, 2020) whose origins can be traced to the anthropogenic deterioration of the global ecosystems.

Morens and Fauci suggested that an ecological framework is needed to understand the COVID-19 pandemic. Its origins, as well as likely future epidemic and pandemic outbreaks, are deeply linked to the exploitation of natural resources and subsequent environmental change and degradation. «Emerging and re-emerging infectious diseases are epiphenomena of human existence and our interactions with each other and with nature. As human societies grow in size and complexity, we create an endless variety of opportunities for genetically unstable infectious agents to emerge into the unfilled ecologic niches we continue to create. There is nothing new about this situation, except that we now live in a human-dominated world in which our increasingly extreme alterations of the environment induce increasingly extreme backlashes from nature» (Morens & Fauci, 2020).

These unbalanced interactions with the natural world are now proving maladaptive also from an evolutionary point of view. Protecting nature is our evolutionary interest. Avoiding the worst consequences of this slippery slope requires farsightedness and a radical rethinking of our development model. The goal we should pursue is a more peaceful coexistence with other inhabitants of the Earth. This is also a necessary step in order to protect ourselves (and our descendants) from an increasingly risky future, characterised, among other things, by ever more frequent pandemics and climatic disasters. This is an evolutionary perspective on what we call One Health.

For this reason, we argue that a scientifically informed "evolutionary responsibility" (Pievani & Lanting, 2019) is needed. As we previously stated, it is impossible to adopt a completely non-anthropocentric perspective. Nonetheless, our unique ability to learn enables us to understand human position in the biosphere and to modify our behaviour, downgrading our claims of superiority and recognising our kinship with other living species. More broadly, we need a humanistic ecological approach which recognises the extent of the impact of human niche construction activities and, at the same time, strives for a more just interaction between humans and nature, as well as within human societies (Belardinelli, 2022).

Geoethics can offer such an ethical framework by assuming a prospective and self-aware weak anthropocentric approach—a "disenchanted anthropocentrism". «At the root of geoethics is the idea that there is a unique community of life on Earth, of which humankind is an inseparable part. Earth is humankind's home, on which our life and future depend. Therefore, as humans, we must respect the Earth and its natural systems and pay attention to how we interact with it, with the awareness that being part of this community involves considering the prudent use of its resources and the conservation of its ecosystems» (Peppoloni et al., 2019). Human embeddedness in nature can be fully understood and accepted from this standpoint, leading to a less anthropocentric (albeit counterintuitive) anthropocentrism. Although not essential to nature, human beings are a legitimate part of it. Our species can therefore pursue its interests, as long as these do not harm other species, which are our neighbours within the biosphere. A weak disenchanted anthropocentrism allows us to recognise both our vulnerability and our power. Consequently, humility and responsibility are the virtues to be cultivated in the Anthropocene.

While humility requires downsizing our claims to dominate nature and reducing our impact accordingly, responsibility, on the contrary, calls for an active commitment to preserve the right to live of our descendants and other living beings. In fact, in the face of the current climatic and environmental crises, adaptation and resilience are not sufficient. We humans have become a geological and evolutionary force on Earth; this requires a full ethical and political commitment, aiming at determining positive consequences even at the evolutionary level. Our learning capacities can drive us to abandon the maladaptive path we are currently in and to build a novel alliance with the rest of the natural world. The needs of human societies and the well-being of ecosystems are not necessarily opposed to each other. However, before implementing any political intervention or technological solution, a theoretical shift is needed.

We believe that cooperation between scientific and humanistic disciplines can drive this transition. Geoethics, enriched by the core concepts of evolutionary sciences and embracing the weakly anthropocentric perspective here described, can thus play a pivotal role in overcoming the present challenges.

Charles Darwin, the father of evolutionary biology, had already envisioned this progressive extension of the "moral circle" in *The Descent of Man* (1871): «As man gradually advanced in intellectual power and was enabled to trace the more remote consequences of his actions […], his sympathies became more tender and widely diffused, extending to men of all races […], and finally to the lower animals—so would the standard of his morality rise higher and higher». This is what is required in order to overcome the current crises and to thrive in the Anthropocene: a progressive extension of the moral circle, the moral recognition of all living beings as equally worthy of consideration and protection.

# References

Adger, W., Barnett, J., Brown, K. N., et al. (2013). Cultural dimensions of climate change impacts and adaptation. *Nature Climate Change, 3*(2), 112–117. https://doi.org/10.1038/nclimate1666

Belardinelli, S. (2022). The human-nature relationship in the anthropocene: A science-based philosophical perspective. *Azimuth, 19*, 19–33.

Crutzen, P. J., & Stoermer, E. F. (2000). *The 'Anthropocene'. Global Change NewsLetter*, Vol. 41. https://inters.org/files/crutzenstoermer2000.pdf. Accessed September 21, 2022.

Darwin, C. R. (1871). *The descent of man, and selection in relation to sex (John Murray).*

Darwin, C.R. (1981). The Descent of Man, and Selection in Relation to Sex. First edition 1871. Princeton University Press.

Dasgupta, P. (2021). *The Economics of biodiversity: The Dasgupta review.* Final Report. HM Treasury. https://www.gov.uk/government/publications/final-report-the-economics-of-biodiversity-the-dasgupta-review. Accessed September 21, 2022.

Di Capua, G., Peppoloni, S., & Bobrowsky, P. T. (2017). The Cape Town statement on geoethics. *Annals of Geophysics, 60*, Fast Track 7. https://doi.org/10.4401/ag-7553

Dirzo, R., Young, H. S., Galetti, M., et al. (2014). Defaunation in the anthropocene. *Science, 345*(6195), 401–406. https://doi.org/10.1126/science.1251817

Elhacham, E., Ben-Uri, L., Grozovski, J., et al. (2020). Global human-made mass exceeds all living biomass. *Nature, 588*(7838), 442–444. https://doi.org/10.1038/s41586-020-3010-5

Ellis, E. C. (2018). Anthropocene: A very short introduction. *Oxford University Press.* https://doi.org/10.1093/envhis/emz017

Ellis, E. C., Gauthier, N., & Goldewijk, K. K., et al. (2021). People have shaped most of terrestrial nature for at least 12,000 years. *Proceedings of the national academy of sciences, 118*(17). https://doi.org/10.1073/pnas.2023483118

Fulgione, D., & Troiano, C. (2019). Clima, abbandono ed evoluzione del paesaggio Mediterraneo (Climate, Abandonment and Evolution in the Mediterranean Landscape). Ambiente Rischio Comunicazione, 16, 65–69 (in Italian).

Gilbert, S. F., & Epel, D. (2015). *Ecological developmental biology: The environmental regulation of development, health, and evolution* (2nd ed.). Sinauer Associates.

Gorenflo, L. J., Romaine, S., Mittermeier, R. A., et al. (2012). Co-occurrence of linguistic and biological diversity in biodiversity hotspots and high biodiversity wilderness areas. *Proceedings of the National Academy of Sciences, 109*(21), 8032–8037. https://doi.org/10.1073/pnas.1117511109

Gould, S. J. (1989). *Wonderful life: The Burgess Shale and the nature of history*. Norton.

Hawking, S. W. (1988). *A brief history of time: From the big bang to black holes*. Bantam Books.

Laland, K., Matthews, B., & Feldman, M. W. (2016). An introduction to niche construction theory. *Evolutionary Ecology, 30*(2), 191–202. https://doi.org/10.1007/s10682-016-9821-z

Levins, R., & Lewontin R. (1985). *The dialectical biologist*. In C. Barner-Barry (Ed.), (Vol. 8). Harvard University Press.

Loh, J., & Harmon, D. (2005). A global index of biocultural diversity. *Ecological Indicators, 5*(3), 231–241. https://doi.org/10.1016/j.ecolind.2005.02.005

Low, F. M., Gluckman, P. D., Hanson, M. A., et al. (2019). Niche modification, human cultural evolution and the anthropocene. *Trends in Ecology & Evolution, 34*(10), 883–885. https://doi.org/10.1016/j.tree.2019.07.005

Maffi, L. (2018). Biocultural diversity. In *The international encyclopedia of Anthropology* (pp. 1–14). American Cancer Society. https://doi.org/10.1002/9781118924396.wbiea1797.

McNeill, J. R., & Engelke, P. (2014). *The great acceleration: An Environmental history of the anthropocene since 1945*. Harvard University Press. https://www.jstor.org/stable/j.ctvjf9wcc

Meneganzin, A., Pievani, T., & Caserini, S. (2020). Anthropogenic climate change as a monumental niche construction process: Background and philosophical aspects. *Biology & Philosophy, 35*(4), 38. https://doi.org/10.1007/s10539-020-09754-2

Morens, D. M., & Fauci, A. S. (2020). Emerging pandemic diseases: How we got to COVID-19. *Cell, 182*(5), 1077–1092. https://doi.org/10.1016/j.cell.2020.08.021

Naudé, P. (2017). Can we overcome the anthropocentrism bias in sustainability discourse? *African Journal of Business Ethics, 11*(2), Article 2. https://doi.org/10.15249/11-2-189

Odling-Smee, J., & Laland, K. N. (2011). Ecological inheritance and cultural inheritance: What are they and how do they differ? *Biological Theory, 6*(3), 220–230. https://doi.org/10.1007/s13752-012-0030-x

Peppoloni, S., Bilham, N., & Di Capua, G. (2019). Contemporary Geoethics Within the Geosciences. In M. Bohle (Ed.), *Exploring geoethics: Ethical implications, societal contexts, and professional obligations of the geosciences* (pp. 25–70). Springer International Publishing. https://doi.org/10.1007/978-3-030-12010-8_2

Peppoloni, S., & Di Capua, G. (2015). The meaning of geoethics. In M. Wyss, & S. Peppoloni (Eds.), Geoethics: Ethical challenges and case studies in earth sciences (pp. 3–14). Elsevier. https://doi.org/10.1016/B978-0-12-799935-7.00001-0

Peppoloni, S., & Di Capua, G. (2021a). Current definition and vision of geoethics. In M. Bohle, & M. Marone (Eds.), *Geo-societal narratives: Contextualising geosciences* (pp. 17–28). Springer International Publishing. https://doi.org/10.1007/978-3-030-79028-8_2

Peppoloni, S., & Di Capua, G. (2021b). Geoethics: An ethics for the relationship between humans and the Earth. *Future of Science and Ethics, 6*, 42–53.

Peppoloni, S., & Di Capua, G. (2021c). Geoethics as global ethics to face grand challenges for humanity. *Geological Society, London, Special Publications, 508*(1), 13–29. https://doi.org/10.1144/SP508-2020-146

Peppoloni, S., & Di Capua, G. (2022a). Crisi ecologica e geoetica. MicroMega, Almanacco di Filosofia(3), 98–108.

Peppoloni, S., & Di Capua, G. (2022b). Geoethics: Manifesto for an ethics of responsibility towards the Earth. *Springer International Publishing.* https://doi.org/10.1007/978-3-030-98044-3

Pievani, T. (2015). Humans' place in geophysics: Understanding the Vertigo of deep time. In M. Wyss, & S. Peppoloni (Eds.), *Geoethics: Ethical challenges and case studies in Earth sciences* (pp. 57–67). Elsevier. https://doi.org/10.1016/B978-0-12-799935-7.00006-X

Pievani, T., & Lanting, F. (2019). La terra dopo di noi. Contrasto.

Pievani, T., & Meneganzin, A. (2020). Homo sapiens: The first self-endangered species. In A. C. Roque, C. Brito, & C. Veracini (Eds.), *Peoples, nature and environments. Learning to live together* (pp. 10–24). Cambridge Scholars Publishing.

Sullivan, A. P., Bird, D. W., & Perry, G. H. (2017). Human behaviour as a long-term ecological driver of non-human evolution. *Nature Ecology & Evolution, 1*(3), 1–11. https://doi.org/10.1038/s41559-016-0065

Troiano, C., Buglione, M., Petrelli, S., et al. (2021). Traditional free-ranging livestock farming as a management strategy for biological and cultural landscape diversity: A case from the Southern Apennines. *Land, 10*(9), 957. https://doi.org/10.3390/land10090957

Worm, B., & Paine, R. T. (2016). Humans as hyperkeystone species. *Trends in Ecology & Evolution 31,*(8), 600–607 S0169534716300659. https://doi.org/10.1016/j.tree.2016.05.008

Xu, C., Kohler, T.A., Lenton, T.M., et al. (2020). Future of the human climate niche. *Proceedings of the National Academy of Sciences, 117*(21), 11350–11355. https://doi.org/10.1073/pnas.1910114117

# Chapter 6
# Materialities, Perceptions and Ethics

Harold P. Sjursen and Luiz Oosterbeek

**Abstract** Ethics is often the distillation of cultural and social tradition, including religious beliefs. Prominent cultural differences are frequently due to geographic and environmental influences which in turn may be reflected in ethical norms. Materialities interplay with understandings and beliefs, since human adaptive strategies are conditioned by the first and addressed by the second, these being guided by foresight and its uncertain drivers, these, in turn, raising the need for ethical considerations. In the West since Biblical times and Greek antiquity the notion that natural resources were a gift to be, used and enjoyed, has been determinative. The emergence of such an understanding may be traced back to water management strategies in given areas of very productive narrow riverine land amid a dominant low natural productivity in most of the territory. Generally, because of the belief that nature was self-renewing and her resources essentially inexhaustible, discussions of the human exploitation of the natural environment have been framed with little acknowledgment of any ethical duties toward nature. This lacuna is challenged in the twentieth century by philosophers such as Martin Heidegger and Han Jonas, and it has subsequently become a prominent discussion among environmentalist philosophers, in an age when the impact of human activity upon the earth has made the need for a global ethics compatible with the geosciences a matter of urgency. This chapter will consider the imperative of responsibility that the new found powers inherent in techno-science have put before humanity and how this duty can be honored.

**Keywords** Technology · Understanding · Responsibility · Ecological · Geoethics

H. P. Sjursen
New York University, New York, USA
e-mail: harold.sjursen@nyu.edu

L. Oosterbeek (✉)
Geosciences Centre, Instituto Politécnico de Tomar, Tomar, Portugal
e-mail: loost@ipt.pt

International Council for Philosophy and Human Sciences–CIPSH, Tomar, Portugal

## 6.1   Introduction

We are becoming aware that the earth, home and support for life as we understand it, is not likely for much longer to sustain the patterns of climate regularity which have allowed part of humanity to flourish with relative ease amidst material comfort. Considering this reality from the singular perspective of humanity (what else can we do?), if anxiety doesn't overcome us, we may experience a form of cognitive dissonance and interpret the growing body of evidence as indications of something entirely different; or we might turn to our creation myths and stories and await divine intervention of one sort or another, (religious or not—Schuman et al., 2018), thus expecting the end of an era, if not of the species itself (like in past millenarian approaches—Skrimshire, 2019); or, in what could be the spirit of rationality, we might invest our hope in the power of techno-science to solve the problem (as a sort of re-enacted positivism—Fawzy et al., 2020). There are numerous variations on these three paradigmatic themes, but no one of them squarely faces the question of ethical responsibility, which is a fundamental one to integrate the material and intangible dimensions of the process, allowing for behavior adaptation and cultural transformation.

In the past, tradition invoked a sense of responsibility through a call or message from the divine; our responsibility was to heed the call. In our era we are presented with no such call. The philosopher Hans Jonas has put it in these terms (Jonas, 1996):

> It was once religion which told us that we are all sinners, because of original sin. It is now the ecology of our planet which pronounces us all to be sinners because of the excessive exploits of human inventiveness. It was once religion which threatened us with a last judgment at the end of days. It is now our tortured planet which predicts the arrival of such a day without any heavenly intervention. The latest revelation—from no Mount Sinai, from no Mount of the Sermon, from no Bo (tree of Buddha)—is the outcry of mute things themselves that we must heed by curbing our powers over creation, lest we perish together on a wasteland of what was creation.

Ethical responsibility is a concept that requires elaboration. Hart (1994) distinguished 4 or 5 senses of the term as used in legal discourse; the philosopher Feinberg (1984) had a lengthier list with a greater emphasis on moral responsibility. But outside of specific interactions between individuals or groups a broader sense of ethical responsibility is difficult to define (Golding, 1986). It would most obviously fit within deontological ethics as a duty. But what kind of duty and what basis that duty would have requires further reflection. Concepts such as "*the good*" and "*duty*" have in recent times been discussed in terms of "*right*" (Rawls, 1971). In the 1970's, Stone (1972), a law professor at the University of Southern California, argued for granting trees a legal voice. While this kind of approach may motivate ecological concern, it is only an attempt to extend existing legal and ethical standards beyond their normal range of application and not substantially change the notion of responsibility itself or clarify a broad sense of ecological responsibility. Furthermore, it raises the ethical question of how can rights be separated from duties/responsibilities (Fredman, 2008) and, through that process, be awarded to living species or other materiality that would not be able to commit to responsibilities. Such an address, in any case, echoes a fundamental ethical flaw of the Human Rights Charter[1]: the non-explicit consideration of related duties.

## 6.2  Materialities and Ethical Imperatives

An ethical imperative for our time is presented by the crisis for humanity dependent on the earth. By this crisis is meant the incipient breakdown of the set of natural dynamics that has controlled and sustained the balance that has permitted advanced life forms to flourish and the concomitant transition toward another yet undefined form of dynamic activity. The ethical imperative emerges from the point of view that human life has a specific kind of consciousness (Damasio, 1999), which renders it responsible in a way different than other species. These are the prerequisites for ethics of any sort: the ability to act in ways that make a difference and the reflective awareness of this ability. The ethical question can be stated roughly as follows: Since we, humanity, to some degree (and probably a very great degree), through our actions have disturbed the earth's ecological balance in a way that will substantially change how and even if the earth can support our life and the lives of other species, does humanity have a responsibility to repair the damage?

The ethical question is seconded by the material question: "1) can humans survive if their ecosystems are disrupted to the point when fundamental materialities, as water or sources of proteins, are destroyed or become insufficient?" and "2) do humans have the capacity and resources to repair damage at the sale of ecosystems?" If self-interest (self-care) is a duty, then the answer would appear obviously to be yes. But, of course, self-interest is a very slippery concept. Is it always in our self-interest to maximize benefit? And, if so, for the long term and for our individual selves only, or for our family, tribe or nation? Or even all humankind and into the indefinite future? And if it is anything like the latter, who gets to decide what constitutes benefit, especially since actions that would produce enduring benefits to the earth and her ecosystem are more than likely to lead to a myriad of perceived hardships.

But apart from these difficulties, one may question what it would even mean for humanity to fix the earth; is it our job to maintain and repair the natural environment and perhaps even to design an improved version? Modern versions of the traditional Jewish ethical principle of וע וקית ל מ [tikun olam] have sometimes been extended to the realm of natural ecology (Troster, 2008), but the aspiration to design an improved version of the world or humanity is something inspired by high-tech futurists like Ray Kurzweil (2006) and represents a techno-spirituality that resides at the pinnacle techno-optimism.

As an ethical issue it falls under the techno-science/rationality paradigm. To regard it under the category of religious promise and morality, such as a literal belief in God's promise recorded in the Bible in the book of Genesis to Noah after the flood never again to destroy the world, steps back from ethical responsibility (Sjursen, 2018).

---

[1] https://www.un.org/en/udhrbook/pdf/udhr_booklet_en_web.pdf (accessed 21 September 2022).

But seeing it within the techno-science paradigm, for example, as an anticipation that a technological fix will alleviate the threats of climate change, raises a host of issues concerning the possibility of ethical responsibility itself in an age dominated by technology (Jonas, 1972). We will return to this issue.

The philosopher Toby Ord's recent book, *The Precipice: Existential Risk and The Future of Humanity*, places humanity of a precipice that is both situational and existential (Ord, 2020). He characterizes the current state of the world as teetering on a precipice where the likelihood is that we fall into an abyss that ends humanity. The reason for this is our own doing; the existential threats of nuclear and biological weapons follow from the progression of technological development from agriculture to our present state of hyper-urbanization. The circumstances we now find ourselves in do not constitute a predicament we can escape from, despite being the circumstantial consequences of the game we have played, that is, it is not a problem to be solved by means of further technology. This very unsettling book, written in consultation with numerous specialists and experts, makes the case that we are now in a genuinely unprecedented state whereby the possibility and perhaps the likelihood that if conditions do not render all life impossible then at least they may lead to the annihilation of human life by degrading our planet to the point where it will be unsustainable. This situation would result from a combination of missteps and negligence, all the consequences of the prevailing capitalist-consumerist-nationalist ways in which humanity is now organized. This goes well beyond a predicament because it strikes directly at the heart of human existence itself. What is meant here?

Past societies did face, in several occasions, the challenge of major climatic changes. While in many cases adaptation entailed primarily migration, i.e., a response consisting of keeping the same behavior pattern while moving away from the area or region where it was no longer sustainable, on occasions it also incorporated a dimension of behavior transformation, as when modern humans adapted to harsher environmental constraints of the late glacial period (McLaughlin et al., 2021), or when they moved from an extensive economy of hunting large mammals, toward a model of economic intensification, through sedentism and domestication (Roffet-Salque et al., 2018). Also, failing to take into consideration the main material drivers of the ecosystem equilibrium led, in the past, to the collapse of countless cultures and civilizations (e.g., in Easter Island—Bahn & Flenley, 1992), while cycles of warming and cooling trends tend to trigger different responses (Oosterbeek, 2022).

## 6.3   Perceptions and Perspectives

Because, as Ord (2020) explains, the *predicament*, even to be understood, must be viewed from the perspective of humanity, a different mode of reflection is required. Ethics, the set of principles and guideposts to help us do the right thing, is most commonly addressed from the perspective of an individual, infrequently from that of a group and recently, but still infrequently, from a global point of view. Even these latter two approaches to ethics do not help because they do not change the perspective, but only zoom out. The perspective question is related to the predicament issue in the context of history. Ord looks at human history as an anthropologist or geologist would, that is, all told it is only a minute fraction of the entire history of the earth. In this way the question of the future viability of human life is that of adjustment and adaptation. Humans were, as they emerged from Homo erectus, at a serious disadvantage compared with other species, not particularly well suited to meet the demands of survival. But the strong cognitive abilities of humanity were put to good use—for example, the making of tools, extending to almost endless limits the innate biological capacities of the human body, or the invention of agriculture which meant that nutritional needs could be reliably met with much less effort and stress. Throughout human history it was through technological innovations on a grand scale such as this that enhanced the quality of life for humans, however generally with an accompanying degradation to the natural environment, which for long was not so severe, due to the low numbers of humans (except in extreme cases, like islands).

This long-standing *game*, according to Ord, has, as it were, largely played itself out, as the environmentalist Bill McKibben (2019) has recently suggested or as Holly Jean Buck (2019) expresses it: "are we at the point—let us call it "the shift"—where it is worth talking about more radical or extreme measures?" Her argument is that we are invested with the notion that nature is stable and self-renewing and consequently averse to the effort to correct the developing imbalance through geo-engineering. At some point the crisis will leave us with no other apparent choice and the pressure of crisis will compel the application of some form of geo-engineering. This is hardly an adequate ethical basis for the shift from regarding the earth as our resource to exploit to re-engineering and re-designing it to suit our perceived needs.

The challenge to such dire and pessimistic assessments calls for a new formulation of the fundamental relationship of humanity to the larger natural order. In the words of Luís Loures (Ergen & Loures, 2021):

> This results in a new attitude towards the environment, which is not a utopic return to the past but, instead, the attitude of identifying the meaning of a sustainable development as a process of change, in which resource utilization, investment, technological development, and institutional changes are in a reciprocal harmony, increasing current and future potential of satisfying human needs and aspirations.

The human game, if by that we mean the constellation of customs, practices and rules that comprise the paradigms of human life on earth, has denied the multiple symbiotic relationships inherent in nature. Even if traditional, and probably prehistoric, communities had a more life-integrated approach to landscape management, as shown in the management of the Amazon rain forest by indigenous communities for the last 5000 years (Piperno et al., 2021), we have come to define life in a manner that ignores or denies constraints on the willed prerogatives of *Homo faber*. Contemporary technology has vastly expanded the belief that human beings are able to control their fate and their environment as a result of the use of tools. This problem is a central theme in Arendt's (1998) classic *The Human Condition*, where she argues that the realm of action, that is, deliberation and choice by an ever renewing and pluralistic community, was usurped (already by Plato) in favor of a politics which favored the artificing of an idealized *polis*.

## 6.4   A Non-anthropocentric Human Perspective

In this chapter we explore the idea of a non-anthropocentric approach to geoethics, moving beyond such artificing approach. The common features shared by all living things, that is, what humanity shares with all life rather than what sets it apart is a likely point of departure. The ethical concerns for the earth's precarious ecosystem, since it affects more than human endeavor, are better understood from a platform that represents life rather than humanity. Such an approach abandons the presupposition that humanity is the apotheosis and perfection of life and deserving of special status because of that. This does not mean, however, that we can understand things as other life forms do, nor that we need to denigrate the status of humanity. Indeed, it is just that which creates the need for human responsibility in the first place. Yet this approach raises questions for which universally agreed upon answers still evade us. The very idea of life is in this category. Hans Jonas (2001) in the *Foreword* to his collection of essays *The Phenomenon of Life* states:

> *"... seek to break through the anthropocentric confines of idealist and existentialist philosophy as well as through the materialist confines of natural science. In the mystery of the living body both poles are in fact integrated. The great contradictions which man discovers in himself —freedom and necessity, autonomy and dependence, self and world, relation and isolation, creativity and mortality— have their rudimentary traces in even the most primitive forms of life, each precariously balanced between being and not-being, and each already endowed with an internal horizon of "transcendence.""*

It is this set of "rudimentary traces" that can provide a point of departure for the kind of non-anthropomorphic ethics we seek as it can offer an interpretation of ethical responsibility. It implies an understanding of life building from its materiality, beyond anthropocentric interpretations of its meaning or function, and in this sense, it also seats apart from current concepts that still perceive humans has deserving a special, even if negative, status (such as the definition of a geological era based on human presence, like the former Anthropozoic, or human action, like the current Anthropocene—Keulartz, 2012). This also supports Jonas's assertion of the impotence of traditional ethical theory in the wake of radical technological transformation.

Geoethics (Peppoloni & Di Capua, 2022 and this volume) offers the possibility of a different perspective than that of contemporary techno-science on humanity's relationship to earth, a perspective from which a mutual dependency and intimacy is perceived. Understanding earlier, pre-scientific and pre-technological, attitudes toward the earth requires historical, ethnographic and hermeneutical studies. The geoethics techno-scientific paradigm needs to include evidence from research from these approaches as well from types of physical evidence. To some extent, this is already incorporated in common practice of geological and archeological investigation; the discovery of tools or utensils in an archeological find, for example, raises ethnographic questions as to how they were used. But if these questions are posed only from the standpoint of utility a modern technological bias is introduced that may occlude understanding other values, as Vere Gordon Childe (1956) suggested long ago.

*The good archaeologist who can himself detach such a flake from the core is in truth re-enacting in his own mind the thought of Mousterian man. He may not be able to express it in an equation and it is certain that he cannot formulate it precisely as the Mousterian would. The latter's rule would probably run something like this: "To make a D-scraper collect a flint nodule (1) at full moon, (2) after fasting all day, (3) address him politely with "words of power," (…)."*

To put it simply, geoethics needs to connect with the past and modern technology may inhibit this. The integrated geoarcheological approach offers a possible model for the kind of multi-disciplinary investigation necessary to approach the ethical problems posed by the present state of the earth.

The philosopher and psychiatrist Thomas Fuchs in an interview about his book defending human being (Fuchs, 2021) offered this observation (Seralathan & Brahmee, 2021):

*Classical humanism is undoubtedly anthropocentric to a high degree, and this can no longer be sustained today. Its lack of consideration of our embeddedness in the earthly environment is all too palpable today in the ecological crisis. The post-humanist criticism of anthro-pocentrism, however, overshoots the mark. To radically question or even want to overcome man because of his misconduct towards nature is absurd—humans are the only beings who can take responsibility for the world, there are no others. As I write in my introduction: Even an ecological redefinition of our relationship with the earthly environment will succeed only if our own embodiment and aliveness—as connectedness or conviviality with our natural environment—is at its centre.*

The issue of how to frame the ethical question presents a genuine ἀπορία (aporia). How would the classical moral virtues of courage, moderation, justice and piety apply to the stance of humanity toward the earth? Our actions toward the earth might exhibit moderation or a kind of justice, but they would be more for our sake than that of the earth. Questions like *what does the earth desire?* or *what gives the earth pleasure?* would be quite misplaced if not absurd. Ethics is about character development and social good, matters pertaining to humanity, individually and collectively. Are we to treat the earth *as though* it were a person and accord it rights and responsibilities as we do with business corporations? This line of inquiry will perhaps only lead us astray.

## 6.5 Ethics Beyond Experiment

Why is it that humanistic studies generally have been ineffective at expressing the realities of climate change in broad existential terms? The novelist Ghosh (2016) has argued in a non-fiction treatise, we live in an era of derangement, that is, of widespread insanity and confusion. In his view phenomena like climate change are traditionally ignored, one reason being simply the limits of human imagination. Historically, whenever societies had to face major climate induces changes, like late hunters in face of the melting glaciers, or Greenland Vikings facing the arrival of Little Ice Age, the major difficulty has been to be able to move from the perceived scale of meteorology to the long-term scale of climate. Maintain a farming economy while the soils got frozen, or thinking meteorology and climate are the same, are expressions of such difficulty, while overcoming it is beyond the scale of experiment, as it requires a degree of abstraction based on probabilistic, science. We could readily agree that the scientific representations of climate change lack existential qualities that may make it seem possible to refute them simply by taking a walk on winter morning. But Ghosh indicts the literary arts for this failure as well. In *The Great Derangement* he explores how the modern novel has been an imaginative failure in the face of global warming. His argument suggests a more general failure, that of the human imagination to deal with the uncanny in any way other than shying away from it. How do we learn to think such phenomena? Hannah Arendt speaks of unlearning and *thinking without banisters* as a necessity in our times (Arendt, 2018). This is the task for the humanities in the face of modern technology. Ghosh's main point, however, is that even in the face of daily evidence of the serious degradation of the earth's environment our larger imaginative faculty does not adequately represent this. A question that underlies the ethical consideration is how can the human imagination assimilate scientific data. This is not a matter of mere scientific literacy, but one of limits of the human brain to assimilate and process all the relevant data, even in the case of scientists.

This ethical question, of whether and if what kind *ethical* responsibility humanity has to repair the damage done to the earth, at once becomes a kind of engineering problem. What would this repair entail? Does it mean to restore the *status quo ante*? Would that even be possible? And if it were, back to when? With this the engineering question reverts to a more purely ethical question. If we could turn back the clock, to what point would that be? To a time when the inhabitants of wealthy nations lived in the comforts produce by industry while most of the underdeveloped world sustained the relative stability of the climate and the abundance of natural resources while their own inhabitants lived lives of far less material comfort? What ethical principle could justify this approach?

In a nutshell, this global debate was performed on occasion of a discussion on the conservation of Lascaux prehistoric cave paintings (Coye, 2011). The cave of Lascaux having been discovered in 1940; it immediately attracted the visit of thousands of people, willing to see the unexpected and very impressive paintings from the Paleolithic. This, however, triggered a degradation of the cave's ecosystem, threatening the preservation of the paintings, and this led to the decision, by the French Minister of Culture André Malraux, of its closure to the public, in 1963, allowing for a relative ecological stability, while a monitoring system was installed. The later corrosion of such device led to its replacement by the turn of the century, and this replacement disturbed once again the ecosystem of the cave, which suffered from a sequence of moisture, fungi and lichen destructive expansion. By 2009, a debate brought together experts from all over the world to discuss on the strategies to face the threat to the paintings: try to retrieve the past ecological balance? Find a new equilibrium? Accept the final degradation of the paintings allowing for their visit? Further restrict visits and preserve the new equilibrium while creating replicas of the cave for the public?

The divide between those (predominantly from earth sciences) that pledged for retrieving past environment and those (predominantly from life sciences and humanities) that considered such an approach impossible and even naïve, as portrayed in the volume edited by Noël Coye, in a sense mimic the current debates on our global responsibilities and possibilities, namely when the discussion is structured around the alternatives of preserving some access to visitors and experts (despite the negative impact on conservation) or further restricting access to all visitors (despite the negative impact on accessibility), even if the final decision that was taken, to produce replicas and further restrict visits to the original cave even to experts, would hardly be acceptable for addressing our global challenges.

Why should a multifaceted ethical question like this take the form of an imperative? On what basis can a duty or obligation to the earth be asserted? Apart from an acknowledgment of a duty to others which recognizes their (or our mutual) dependency for survival upon largess of the earth, in what way are any of us obligated to the earth? What is our duty and to whom or what are we so obliged? To talk about a duty to the earth is something in traditional Western ethics rarely considered. If it is a duty to humanity, then the ethical issue is exacerbated because, especially now given the power of technology, we are often dealing with people we will never meet, the inhabitants of the earth in the indefinite future.

As humans, what are we entitled to from our natural environment? Many would say clean air and pure water. What about biodiversity, temperate climate? And what to say about cultural diversity, itself rooted in diverse ecological contexts, including cultural heritage, as evidenced in the case of the cave of Lascaux? The process of life is one of consuming—how much of that is our right?

These kinds of issues, resistant as they are to our most frequent norms of evaluation and analysis, have largely emerged in an age dominated by technology and are what prompted Hans Jonas to call for a new ethics for a technological age (Jonas, 1979). The points he made in support of this proposal might be summarized as follows:

(1)  Technology has advanced from tool to machine to automatic device—this leads to the situation where some technology may be beyond human control.
(2)  Technological processes are often not well understood and produce unanticipated consequences—emergent technologies manifest a high degree of human ignorance due to complexity.
(3)  Technology produces results disproportionate to human action—this raises the issue of overwhelming power, as we can destroy the world with the simple push of a button.
(4)  Technology may alter the environment permanently—that is, our actions may be irreversible.
(5)  Results of technology may only present themselves in the distant future—because potentially damaging consequences affect the unknown and indefinite future, they lie beyond normal motivations for our concern.

The two main aspects of his thesis are that the nature of human action has changed such that we now possess extraordinary power, entirely disproportionate to our natural stature, capable of destruction beyond the limits of our imagination.

This implies a qualitative change in relation to past human extractive strategies, as those were framed within a space–time scale that allowed for regeneration, even if this could be occasionally challenged, namely after the industrial revolution. The current technological impact operates in a new time scale, beyond the possibility of human perception and of ecosystem resilience, which sets a new framework for human reflection and action. This technologically mediated power, in addition to surpassing our imaginative faculties, is in a sense not controlled by us because its consequences are unpredictable. We are able to initiate processes that we do not fully understand, that are powerful beyond our natural means, and which may yield consequences unanticipated and unperceived by us.

## 6.6   Understanding the Past and Technology

These are the questions that motivate *geoethics*. The motivation is strengthened by the belief, supported by the ever-increasing power of techno-science, that the earth is truly under our dominion, that we can control and manage it to suit our needs and preferences, even when scientists demonstrate this is not the case. Let us review some of the issues that need resolution.

Most recent discussions concerning the ethics of geo-engineering *solutions* to the hardship and destruction to the natural environment by anthropogenic climate change have focused on cost, benefit and risk analysis and whether on those grounds geo-engineering solutions are justified ethically (Mittiga, 2019). This approach is not discussed here, although the notion that a radical intervention designed to alter the natural operations of the globe requires, we believe, greater justification than the results of cost, benefit and risk analysis.

A major driver of historical and anthropological assessment of the past concerns the understanding of *human being/environment* adaptive processes as representing a growing mastering of environmental resources (i.e., raw materials) through by technology, specific techniques combined with logistics. The progress of civilization was measured by such means which, in the face of the unintended consequences of environmental degradation must be viewed in a different light. This kind of assessment is conducted under set of operative intellectual categories, namely the presuppositions of science concerning the natural environment. But this may limit our understanding in a way as suggested by Heidegger and others. Although science as we understand the term descended from the ancient Greeks, technology existed prior to the advent of this mode of understanding (Lloyd, 1970), yet at this stage technology as human craftsmanship did not was not mediated by the theoretical constructs of philosophy or science.

Despite the awareness of scholars of the explanatory limitations of this approach (in terms of the understanding of motivation related to choices, namely given the lack of sufficiently accurate information on intangible cultural and psychological drivers), such an assessment allows one to approach contextual constraints of human behavior and offers examples of sustainable management of the environment in the past, even if edaphology, biome or demographic variables are very different from present ones. There is a consensus among scholars that first written narratives (religious or not) are written later versions of earlier oral, mostly sang, tales, include behavior prescriptions that bridge with current notions of moral and ethics. But we know very little about the mindset of these early societies. The very idea of *nature* may not have been operative, let alone thought about in anything like scientific terms, in ancient Mesopotamian culture, yet they did develop a way of understanding and relating to their natural environment (Rochberg, 2016). Such traditions at least suggest the possibility of an alternative mode, one not wedded to the current prevailing mechanistic model, for understanding our own relationship to our natural environment.

However, the approach of humanity to its past, based alone on the material evidence through a logic of resources management is, in fact, in line with an understanding of the industrial societies of their contexts as being resources alone. In this sense, the assessment of the past, necessarily reduced to the tangible dimension due to the absence of other evidence, tends to lead to a societal approach to the present also along those lines (the notion of progress being embedded in it). This triggers, today, different types of reaction, that may be clustered around two extreme positions: the negation of the interest of history and archeology, replacing them with a literary non-scientific revision of the past, in order to demonstrate past adaptive sustainable behavior patterns (from the imagination about past societies "will" to the speculations around the "Paleolithic diet," which find no hard evidence from archeological research); the negation of the relevance of non-material dimensions for the understanding of past societies (and, as a result, of contemporary societies, explained through the notions of competition, survival of the fittest).

This division, in attitudes toward nature and the role of human activity, generated two disparate camps, one a limited techno-scientific view and the other a nostalgic understanding derived from literary accounts produced by a truncated imagination.

The assessment of the past entails, hence, wider ethical implications, concerning the legitimacy of given approaches, their validity across time and beyond specific contexts. Also, technology and logistics being core drivers of the academic study of the past (and of ethology studies involving other species too), bringing in ethical concerns that build form those axes may prove to be useful to embrace a more diverse understanding of human behavior without falling into negationism and anti-science. Yet, without abandoning the fruit of these inherited practices, geoethics needs to assimilate some of the insights offered by traditional perspectives.

One might see geoethics as it is being proposed here in light of the overview provided already in the 1930s by Lewis Mumford in his call-to-action for the humanity to consider its options in the face of the threat to its very survival by ecological catastrophe or industrialized warfare (Mumford, 1934). At about the same time Heidegger was formulating a critique of technology which became more specifically a negative critique of cyber-technology. His analysis was derived from his reading of the meaning of technology in classical Greek philosophy. His argument when put together with the notion of *Gaia* suggests an interesting approach to a kind of geoethics that is less influenced by techno-science and which steps back from a purely "resources and use for humans" perspective. Let us consider some of the key aspects of this perspective.

Heideger's later (post *Zein und Zeit*) perspective (Heidegger & Stambaugh, 1996), when combined with aspects of Santiago theory and the Gaia worldview (Dicks, 2011), suggests an ethical program. Heidegger's position rests upon his understanding of ποίησις (poiesis) which in classical Greek thought means the activity by which something that did not exist is brought into being. It is close to τέχνη (tekhne), making or doing, and πρᾶξις (praxis) which refers to the activity or practice of rendering an idea or skill as action. Poiesis aims at a product whereas for praxis the end is action. Poieisis in Aristotle's account is what artists and poets do, thus changing the characterization of the work of art from imitation as Plato maintained to something new brought forth. It is the concept of bringing forth that Heidegger concentrates on. In the Greek view tekhne covered both handicraft and artistic production such as poetry. But, according to Heidegger's understanding, the process of bringing forth (poiesis) has two senses: in the first what is brought forth is brought forth by something else as when the poet brings forth the poem; in the second sense poiesis is physis (nature), i.e., the bringing forth that occurs in nature, like the bringing forth of a plant (Heidegger, 1977).

Heidegger argues that techno-science understood by him as dominated by cybernetics, conceals or hides the fullness of Being and thus prevents philosophy (us) from thinking Being and thereby leads us to apprehend Being merely in terms of resource or standing reserve (*Bestand*).

The idea of *autopoiesis* is an important concept in biological theory, referring self-organizing/self-generating systems, e.g., single-cell animals like the amoeba, or more generally cellular life forms. It has been argued that autopoiesis is the defining principle of life itself. This idea comes from the work of Maturana and Varela (1980), who coined the term. Most controversially was their claim that this organizing activity (on the cellular level) was the equivalent of cognition.

The applicability of Santiago Theory, as the work of Maturana and Varela is known, is taken to establish a new paradigm in biology by the physicist turned biologist Fritjof Capra who emphasizes a systems approach as a corrective for the mechanistic approach (Capra & Luisi, 2014). According to Capra, nature generally has been understood by modern science in what he calls the Newtonian mechanistic worldview in which the cosmos is seen as a kind of machine, an approach concordant with (although not demanded by) today's techno-science.

Technology has been an important issue since Greek philosophy distinguished and divided *episteme* from *technics*, associating the latter with the rhetoric of sophistry. This traditional distinction and its pejorative association have influenced the development of European civilization and helped to shape the discourse of art, religion, politics, science and much else. The scientific revolution of the seventeenth century and the subsequent debates surrounding Darwin's evolutionary biology expressed aspects of the tension between technics as mechanics versus the science of beings that possessed the qualities of agency, self-organization and purpose. These debates revived the disputes around Descartes' dualism in which the human body was a mechanical device with no agency of its own.

## 6.7 Understanding Symbiotic Relations

The mechanistic view of animal life, consistent with Cartesian dualism, applied to human bodies but not to the human mind–body aggregate. However problematic this dualistic conception of being human was (how could utterly unlike substances—*res cogitans* and *res extensa*—manage to interact?), it served the emerging *new science* well. Human agency somehow derived from cognition understood as a process absent from creatures and life forms save the human being whose rationality was now appended to animality. Darwin whose evolutionary biology of random selection would fit a mechanistic interpretation himself had a less than mechanistic understanding of animal behavior and choice. About animals he opined: "the exertion of choice, the influence of love and jealousy, and the appreciation of the beautiful in sound, colour or form" was often determinative (Darwin, 1981).

Capra mentions early Greek *hylozoism* with an appreciation that suggests that he thinks that the replacement of that worldview by the radical separation of living from non-living matter, necessary for the mechanistic theory, later contributed to the rise of that theory and the doctrine of non-teleological explanation for change and action in the domain of the material. He argues that a systems or network approach, on the contrary, permits a non-mechanistic understanding of the human body and other levels of life (notably the molecular). The problem of the apparent agency of living things, in this approach together with the concept of autopoiesis, is thereby solved (Capra & Luisi, 2014).

The meaning of autopoiesis for a theory of life is summarized by Bitbol and Luisi (2004) as follows:

> The theory of autopoiesis, as developed by Maturana and Varela, ... captures the essence of cellular life by recognizing that life is a cyclic process that produces the components that in turn self-organize in the process itself, and all within a boundary of its own making. The authors thus arrived at the definition of an autopoietic unit, as a system that is capable of self-maintenance owing to a process of components self-generation from within. This generalizes the definition of life. Systems involving RNA-DNA coding (as in actual cells) are no longer the only possible living entities. The important notion is that the activity leading to life is a process from within, i.e. dictated by the internal system's organization. This 'activity from within' permeates all other concepts associated to autopoiesis, like the notion of autonomy, or biological evolution, or the rules of internal closure.

When the autopoietic interpretation is supplemented by the Gaia hypothesis a new basis for geoethics is presented.

The Gaia Hypothesis proposed by James Lovelock (1995) suggests that living organisms on the planet interact with their surrounding inorganic environment to form a synergetic and self regulating system that created, and now maintains, the climate and biochemical conditions that make life on earth possible. Gaia bases this postulate on the fact that the biosphere, and the evolution or organisms, affects the stability of global temperature, salinity of seawater and other environmental variables. For instance, even though the luminosity of the sun, the earth's heat source, has increased about 30% since life began almost four billion years ago, the living system has reacted as a whole to maintain temperatures at a level suitable for life.

Cloud formation over the open ocean is almost entirely a function of oceanic algae that emit sulfur molecules as waste metabolites which become condensation nuclei for rain. Clouds, in turn, help regulate surface temperatures.

Lovelock compared the atmospheres of Mars and earth and noted that the earth's high levels of oxygen and nitrogen were abnormal and thermodynamically in disequilibrium. The 21% oxygen content of the atmosphere is an obvious consequence of living organisms, and the levels of other gases, $NH_3$ and $CH_4$, are higher than would be expected for an oxygen-rich atmosphere. Biological activity also explains why the atmosphere is not mainly $CO_2$ and why the oceans are not more saline. Gaia postulates that conditions on earth are so unusual that they could only result from the activity of the biosphere (Reichle, 2020).

The relationship between living organisms and their inorganic environment is symbiotic, with benefit flowing in both directions, and not one of mere dependency. For instance, the archeological interpretation of the domestication of living species in the dawn of farming, occurring independently in different regions of the planet in the Holocene, became an adaptive possibility because, even before, late hunter-gatherers had for long engaged in a process of specialization that allowed not only to better understand their reproductive and habitat strategies, but also to curate these. The process of selection of species, benefiting some to the expense of others, also became expression of a symbiotic relation between humans and animals or plants, as Eric Higgs demonstrated long ago (Higgs & Jarman, 1969). More recently, studies on the management of the Amazon rain forest by indigenous communities for the last 5000 years are another example of such symbiotic relation (Posey et al., 2006).

Within such an understanding, the living/non-living environment is a whole, systemically ordered, where humanity is an integral part. Moreover, from this perspective the living-non-living dichotomy has less significance. Yet, although Lovelock (1995) describes Gaia in affectionate terms, at times comparing human emotions and feelings to something inherent in earth, from the human perspective it has not always looked that way. From the human point of view, and hence from the viewpoint of traditional ethics, humans stood in a unique and privileged relation to the earth or non-human nature. The characterization of earth as our home, while it supports an ethic of care, does little to challenge this structure. For this reason, the Gaia hypothesis offers the possibility of a corrective to an ethics proscribed by classical metaphysical categories, whether Aristotelian or Cartesian.

## 6.8   Understanding of Life

We will now return to Heidegger whose position, on this issue, like those of Hans Jonas and Toby Ord, expresses an existential sense of crisis. There is also a sense in which theories about the so-called post-human era suggest a similar anxiety. Bruno Latour's Actor Network Theory presents an account of technology—machines and systems—that function in a collaborative fashion in a distributed network. His assessment of the relation of humanity to technology (Latour, 1987) stands in contrast to that of Heidegger.

Heidegger's understanding of truth and the concealing/revealing of it by technology can help guide our understanding of nature. In his 1927 book *Sein und Zeit* (Being and Time) indicted the Western tradition from Plato onward for having misconstrued the basic issue of what it means for something to be, i.e., how something is present to an attentive human mind absent any scientific or philosophical analysis or interpretation. For Heidegger, *forgetting* of Being as he puts it has led us into the crises which characterize modernity. In his later work he argues that the prevailing scientific/technological approach, problem-solving rather than thinking, presents nature as raw material and blinds us to its and our true nature. Heidegger does not propose that we reject technology, but that we recognize its danger. Rather, it is to stand in a free relation to technology, one that implies that the discourse of Being is not obfuscated by the metaphysical categories adopted by cybernetic science.

Does this understanding of life and the earth as a symbiotic and interdependent part of the life system bring us any closer to a new ethics for an age of technology as Hans Jonas put it? Such an ethics cannot eschew technology altogether but by the same token it must not yield to the interpretive power latent in technology use. Aristotle (2011) did not think, in contrast to Plato, that one needed to possess theoretical knowledge in order to know what goodness is. For him it was more of a habituated attitude, one that we might compare in this case to the attitude James Lovelock attributes to his father.

> *He had not formal religious beliefs and did not attend church or chapel. I think his moral system came from that unstructured mixture of Christianity and magic which is common enough among country people, and in which May Day as well as Easter Day is an occasion for ritual and rejoicing. He felt instinctively his kinship with all living things, and I remember how greatly it distressed him to see a tree cut down.*

But while this attitude is surely conducive to personal virtue, the particular problem that high-tech solutions present is a kind of ignorance of outcomes beyond the immediate that is not overcome by an appreciation of nature or the desire to protect the earth. Wildlife conservation is a beautiful ideal that may not grasp the dynamic character of the natural environment considered holistically. It seems that a knowledge of aspects of techno-science is a fundamental necessity for recognizing what one is doing as a powerful agent of environmental change. A new ethics for the technological future—now upon us—must use technology to reveal and disclose the being in which we dwell.

There is another sense in which the problem of what prerequisite knowledge ethics requires is clearly recognized by Aristotle. The pursuit of the good may lead to disagreement as to what is the highest good. This kind of conflict is often evident in disputes about the natural environment: agriculture or mining versus pristine wilderness, for example. The expectation that land can be returned to its *natural* state after it has been fully mined represents both an inadequate understanding of ecosystems and an abundance of faith in the power of technology. The adjudication of this kind of conflict requires not only knowledge of what is actually possible, but also evaluate differing preferences and notions of what is good and beautiful.

In a highly tentative way, we shall suggest that the work of earth scientists can contribute to the development of a kind of geoethics that embraces the affirmation of life as a source of value and contributes to our knowledge of the earth's ecological systems through the reflective use of modern technology.

To try to illustrate how such an approach could expand geoethics to include, in addition to material records and the investigative power of geosciences and geo-engineering, literary reflections, historical records and hermeneutical analytics as part of a comprehensive effort to address the crisis of the earth. We could suggest, for example, that a study of the meaning bestowed upon rivers in ancient societies, particularly in China, would exhibit this kind of integrative attitude. It has been pointed out frequently that the history of China can be told through its persistent efforts to live with its rivers, especially the Yellow River and the Yangtze. These rivers and their profound impact on nearly all aspects of life in ancient China were understood first through the lens of mythology and then by Daoist philosophical inter-pretations. Because of this in ancient China the laws of nature are seen as embedded in moral precepts. To state this in contemporary language, it would be as though earth science was an aspect of ecological ethics—not the reverse which is our usual perspective. To explain this, before looking more specifically at the example of the two great rivers, a precise of Chinese cosmological theory is needed.

In the Song dynasty neo-Confucian rendering of classical Chinese cosmology— which in a sense offers a grand synthesis of classical Confucian, Daoist and Buddhist worldviews in the context of Zhou dynasty cosmology—the distinction between nature and humanity is not operative; there is an interpenetration of (human) mind and cosmos such that each is responsive to the other. The Western metaphysical categories of *self* and *other* are dissolved. The self, heart/mind, is responsive to the natural order as which is its extension. Nothing like the Cartesian dualism is present. The ethics of the *self* is coextensive with the natural order from which it is inseparable (Sjursen, 2015).

The Yellow and the Yangtse Rivers in China have done much to shape Chinese civi-lization and simultaneously inform the Chinese imagination (Ball, 2017). Although the rivers were and are vital to agriculture, transportation and much else that char-acterizes material Chinese culture, they were understood in terms completely other than as "standing reserve." Despite early innovations by the Chinese in hydraulic engineering and water management, the quasi-spiritual regard for the rivers was not lost (Needam, 1971). The vitality and life of the rivers was understood as revealing life and vitality itself. This recognition of connectedness certainly makes possible a form of ecological consciousness in which an ethics of the earth is coupled with a reverence for life and where technology does not distort the intimacy of humanity's participation in nature. The traditional Chinese skepticism toward technology, despite its impressive advancements, seems to have yielded an attitude toward nature, and particularly rivers as living waters, that has inspired cross-cultural understanding and cooperation (He, 2000).

The Daoist sage Zhuangzi both recognized the utility of technology and then how it can impinge upon human relations to nature as illustrated in a story from Chapter 12 which tells of an encounter between Kongzi's disciple Zigong and an old man watering his vegetables. Zigong is traveling in the country when he encounters the old man laboriously hauling water. Being sympathetic, Zigong stops to tell the old man about a machine, the well sweep, that can raise enough water for a hundred fields in the time it takes him to water one. As he listens, the old man begins to grimace, his face reddens, and finally he tells Zigong that he knows all about this machine and would be ashamed to use it. The old man is not unlearned. He mentions a master, who taught him that ingenious machines (ji xie 機械) require ingenious minds (ji xin 機心); that ingenious minds cannot be pure and simple (chun bai 純白); that they are restless and no restless mind (shen sheng bu ding 神生不定) moves with dao. That said, he pronounces a curse on scholars and another on Zigong and returns to his work (Zhuangzi, 2003).

The story can be read in Heideggerian terms: not as an outright rejection of technology, nor even of technology used to access the earth's resources, but as a questioning of how the Being of humanity and that of nature dwell together.

## 6.9 Conclusion

The theme of this chapter has been "Materialities, Perceptions and Ethics." The three terms are linked through the complex relationships between human beings and the earth. Materiality refers to biology and geo-physics; perceptions include belief systems and worldviews; ethics broadly includes actions and behaviors that aim to preserve or promote the good, however that may be understood.

Our argument regarding materialities is that bio- and earth sciences have been hampered by or biased in favor of the canonical mechanistic world view favored by modern physics. The limitations of this view have been discussed by Fritjof Capra who thinks the phenomenon of life is more accurately portrayed when the earth is understood as a network of systems, abandoning the machine analogy (Capra & Luisi, 2014). This view both addresses the problem of agency without the invocation of supernatural intervention and permits for a mutually symbiotic relationship between organisms and their supportive environment. Capra's views line up in many respects with the *Gaia* hypothesis which we have suggested is in the tradition of hylozoistic theories from early Greek physics as well as from other similar traditions (Chinese has been mentioned; there are many relevant discussions in the Indian Vedanta literature) as well as contemporary theories and investigations of the *problem of consciousness*. We have stopped short of endorsing notions of cosmic consciousness (such as those advanced by Richard Bucke (2010) who argued for the reality of cosmic consciousness as "a higher form of consciousness than that possessed by the ordinary man [… one whose …] prime characteristic … is, as its name implies, a consciousness of the cosmos, that is, of the life and order of the universe.") Our task has not been to solve the problem of consciousness as understood by either philosophy of neuroscience, but only to suggest that thinking about the earth and human nature in the categories supplied by Descartes and early modern science makes the task of developing non-anthropocentric ethics that includes a duty to care for the earth beyond reach.

In response to the challenge of Jonas (1972) for a new ethics for an age of technology that can support what he calls the imperative of responsibility, we argue that this requires a non-anthropocentric understanding of life and a non-cybernetic view of nature. Since ethical obligations and principles persist only in cultural settings, anthropology, archeology, esthetics and the study of literature and religions and other disciplines from the humanities and social sciences must investigate collaboratively to identify an approach that honors a pluralistic set of values. While geoethics alone is unlikely to answer Jonas's challenge, because it rests upon in a non-reductionist approach to science (earth sciences including biology, geology and archeology) which itself involves humanistic disciplines including anthropology, history, as well as esthetics (landscape architecture in part concerns itself with this), a geoethics that recognizes in the Heideggerian sense the dangers represented by technological society is the best the place to start.

**Acknowledgements** Luiz Oosterbeek would like to thank the Portuguese Foundation for Science and Technology for its support to research leading to this chapter, through the Geosciences Centre (R&D unit 00073) and the Polytechnic Institute of Tomar.

# References

Arendt, H. (1998). *The human condition.* The University of Chicago Press.

Arendt, H. (2018). Thinking without a Banister: Essays in understanding, 1953–1975. In J. Kohn (Ed.). Schocken ed.

Aristotle. (2011). *Aristotle's Nicomachean Ethics* (R. C. Bartlett, & S. D. Collins, Trans.). University of Chicago Press.

Bahn, P. G., & Flenley, J. (1992). *Easter Island, Earth Island.* Thames and Hudson.

Ball, P. (2017). *The water kingdom: A secret history of China.* The University of Chicago Press.

Bitbol, M., & Luisi, P. L. (2004). Autopoiesis with or without cognition: Defining life at its edge. *Journal of the Royal Society Interface, 1,* 99–107. https://doi.org/10.1098/rsif.2004.0012

Buck, H. J. (2019). *After geoengineering: Climate tragedy, repair, and restoration.* Verso.

Bucke, R. M. (2010). *Cosmic consciousness: A study in the evolution of the human mind.* Martino Publishing.

Capra, F., & Luisi, P. L. (2014). *The systems view of life: A unifying vision.* University Press.

Childe, V. G. (1956). *Piecing together the past: The interpretation of archeological data.* Routledge & Kegan Paul.

Coye, N. (Ed.). (2011). Lascaux et la conservation en milieu souterrain/Lascaux and preservation issues in subterranean environments. Actes du symposium international/Proceedings of the International Symposium Paris, 26 et 27 février 2009/Paris, February 26 and 27, 2009. Paris: Éditions de la Maison des Sciences de l'Homme.

Damasio, A. R. (1999). *The feeling of what happens: Body and emotion in the making of consciousness.* Harcourt Brace & Co.

Darwin, C. (1981). *The descent of man, and selection in relation to sex*. Princeton University Press.

Dicks, H. (2011). The self-poetizing Earth: Heidegger, Santiago Theory, and Gaia Theory. *Environmental Philosophy, 8*(1), 41–61. https://philpapers.org/rec/DICTSE-2

Ergen, M., & Loures, L. (2021). *Landscape architecture: Processes and practices towards sustainable development*. Intechopen.

Fawzy, S., Osman, A. I., Doran, J., & Rooney, D. W. (2020). Strategies for mitigation of climate change: A review. *Environmental Chemistry Letters, 18*(6), 2069–2094. https://doi.org/10.1007/s10311-020-01059-w

Feinberg, J. (1984). *The moral limits of the criminal law: Offense to others*. University Press.

Fredman, S. (2008). *Human rights transformed: Positive rights and positive duties*. Oxford University Press.

Fuchs, T. (2021). *In defence of the human being: Foundational questions of an embodied anthropology*. Oxford University Press.

Ghosh, A. (2016). *The great derangement: Climate change and the unthinkable. In The Randy L and Melvin R Berlin Family Lectures*. The University of Chicago Press.

Golding, M. P. (1986). Issues in responsibility. *Law and Contemporary Problems, 49*(3), 1–12. https://scholarship.law.duke.edu/lcp/vol49/iss3/1/

Hart, H. L. A. (1961) [1994]. *The concept of law*. In P. A. Bulloch, & J. Raz (Ed.), Clarendon Press.

He, M. F. (2000). *A river forever flowing: cross-cultural lives and identities in the multicultural landscape*. Information Age Publishing.

Heidegger, M. (1977). *The question concerning technology, and other essays*. Garland Publishing, Inc.

Heidegger, M. (1996). *Being and time: A translation of Sein und Zeit. Translated by Joan Stambaugh*. State University of New York Press.

Higgs, E. S., & Jarman, M. R. (1969). The origins of agriculture: A reconsideration. *Antiquity, 43*(169), 31–41. https://doi.org/10.1017/S0003598X00039958

Jonas, H. (1972). Technology and responsibility: Reflections on the new task of ethics. In *Religion and the humanizing of man: International Congress of Learned Societies in the Field of Religion* (pp. 1–19), Los Angeles, Calif, Waterloo. https://inters.org/jonas-technology-responsability

Jonas, H. (1979). Das Prinzip Verantwortung: Versuch e. Ethik Für d. Technolog. Zivilisation. Frankfurt am Main: Frankfurt am Main: Insel-Verlag.

Jonas, H. (1996). *Mortality and morality: A search for good after Auschwitz. Edited and with an introduction by Lawrence Vogel*. Northwestern University Press.

Jonas, H. (2001). *The phenomenon of life: Toward a philosophical biology*. Northwestern University Press.

Keulartz, J. (2012). The emergence of enlightened anthropocentrism in ecological resoration. *Nature and Culture, 7*(1), 48–71. https://doi.org/10.3167/nc.2012.070104

Kurzweil, R. (2006). *The singularity is near: When humans transcend biology*. Penguin Books.

Latour, B. (1987). *Science in action: How to follow scientists and engineers through Society*. Harvard University Press.

Lloyd, G. E. R. (1970). *Early Greek science: Thales to Aristotle*. W. W. Norton & Company.

Lovelock, J. (1995). *Gaia: A new look at life on Earth*. Oxford University Press.

Maturana, H. R., & Varela, F. J. (1980). *Autopoiesis and cognition the realization of the living*. Springer, Netherlands.

McKibben, B. (2019). *Falter: Has the human game begun to play itself out?* Henry Holt and Co.

McLaughlin, T. R., Gómez-Puche, M., Cascalheira, J., Bicho, N., & Fernández-Lopez, J. (2021). Late Glacial and Early Holocene human demographic responses to climatic and environmental change in Atlantic Iberia. In *Philosophical transactions of the Royal Society, Series B: 376*. https://doi.org/10.1098/rstb.2019.0724

Mittiga, R. (2019). What's the problem with geo-engineering? *Social Theory and Practice, 45*(3), 471–499. https://doi.org/10.5840/soctheorpract201992768

Mumford, L. (1934). *Technics and civilization*. Harcourt, Brace and company.

Needham, J. (1971). *Science and civilisation in China, Vol. 4: Physics and physical technology, Part 3: Civil engineering and Nautics.* Cambridge University Press.

Ord, T. (2020). *The precipice: Existential risk and the future of humanity.* Hachette Books.

Oosterbeek. (2022, in press). From global warming into a New Ice Age? climate, adaptation and examples from the past. Leading-edge lecture series. Taipei: NTU Institute for advanced studies in the humanities and social sciences.

Peppoloni, S., & Di Capua, G. (2022). Geoethics: Manifesto for an ethics of responsibility towards the Earth. *Springer, Cham.* https://doi.org/10.1007/978-3-030-98044-3

Peppoloni, S., & Di Capua, G. (2023, this volume). Geoethics for Redefining Human-Earth System Nexus.

Piperno, D. R., McMichael, C. H., Pitman, N. C. A., Ernesto Guevara Andino, J., Paredes, M. R., et al. (2021). A 5000-year vegetation and fire history for tierra firme forests in the Medio Putumayo-Algodón watersheds, northeastern Peru. *PNAS, 118*(40), e2022213118. https://doi.org/10.1073/pnas.2022213118

Posey, D. A., Balick, M. J., Balick, M., & Posey, D. (2006). *Human impacts on Amazonia: The role of traditional ecological knowledge in conservation and development.* Columbia University Press.

Rawls, J. (1971). *A theory of justice.* Belknap Press of Harvard University Press.

Reichle, D. E. (2020). Chapter 2—The physical and chemical bases of energy. In: D. E. Reichle (Ed.), *The global carbon cycle and climate change* (pp. 5–14). Elsevier. https://doi.org/10.1016/B978–0–12–820244–9.00002–0

Rochberg, F. (2016). *Before nature: Cuneiform knowledge and the history of science.* University of Chicago Press.

Roffet-Salque, M., Marciniak, A., Valdes, P. J., Pawlowska, K., Pyzel, J., et al. (2018). Evidence for the impact of the 8.2-KyBP climate event on near eastern early farmers. *PNAS, 115*(35), 8705–8709. https://doi.org/10.1073/pnas.1803607115

Schuman, S., Dokken, J.-V., van Niekerk, D., & Loubser, R. A. (2018). Religious beliefs and climate change adaptation: A study of three rural South African communities. *Journal of Disaster Risk Studies, 10*(1), a509. https://doi.org/10.4102/jamba.v10i1.509

Seralathan, P., & Brahmee, B. (2021). Humans are the only beings who can take responsibility for the world: Thomas Fuchs. The Hindu. https://www.thehindu.com/society/humans-are-the-only-beings-who-can-take-responsibility-for-the-world-there-are-no-others-thomas-fuchs/article37002888.ece. Accessed September 21, 2022.

Sjursen, H. P. (2018). Technological ethics, faith and climate control: The misleading rhetoric surrounding the Paris agreement. In V. Popovski (Ed.), *The implementation of the Paris agreement on climate change* (pp. 151–163). Routledge.

Sjursen, H. P. (2015). Zhu Xi's pure practicality: The 'Self' in traditional Chinese thought'. http://harold-sjursen.org/new-page-3. Accessed September 21, 2022.

Skrimshire, S. (2019). Activism for end times: Millenarian belief in an age of climate emergency. *Political Theology, 20*(6), 518–536. https://doi.org/10.1080/1462317X.2019.1637993

Stone, C.D. (1972). Should trees have standing? Toward legal rights for natural objects. Southern California Law Review (Vol. 45, pp. 450–501). https://iseethics.files.wordpress.com/2013/02/stone-christopher-d-should-trees-have-standing.pdf. Accessed September 21, 2022.

Troster, L. (2008). Caretaker or citizen: Hans Jonas, Aldo Leopold, and the development of Jewish environmental ethics. In H. Tirosh-Samuelson, C. Wiese (Eds.), *The legacy of Hans Jonas* (pp. 373–396). Brill. https://doi.org/10.1163/ej.9789004167223.i-578.81

Zhuangzi. (2003). *Zhuangzi: Basic writings.* Transl. Burton Watson. New York: Columbia University Press.

# Chapter 7
# Affirmative Ethics, New Materialism and the Posthuman Convergence

Rosi Braidotti

**Abstract** This chapter addresses the challenge of the posthuman convergence between great technological advances—the Fourth Industrial Revolution—and advanced environmental emergency—the Sixth or even Seventh Extinction. It argues that an ethics of affirmation constitutes a robust alternative to the state of anxiety that marks this era. The chapter outlines the defining features of affirmative ethics in the continental philosophical tradition of new or non-reductive materialism and relational life philosophies. It emphasizes immanence, grounded-ness and a nature-culture continuum, which also encompasses technological mediation as our second nature. The chapter defends the relevance of neo-materialist affirmative ethics, with emphasis on feminist, environmentalist, race and indigenous theories.

**Keywords** Affirmation · Relational ethics · Posthuman thought · Feminist new materialism · Spinoza studies · Indigenous philosophy

## 7.1 Introduction

My argument in this chapter is that new materialism is the most relevant philosophical framework for rethinking the relationship to the complex environmental and social ecologies that define the posthuman era. New materialism foregrounds the need to establish a relational collaboration with the material eco-systems and its non-human entities, but also the social *milieus* and their normative practices, while also inscribing such relational bonds in a framework of technological mediation. This nature-culture-technology continuum reflects the two-pronged framework of the posthuman convergence. Namely, the co-occurrence of advanced, albeit uneven, technological development—the Fourth Industrial Revolution (Schwab, 2015) on the one hand, and of fast environmental depletion—the Sixth (Kolbert, 2014) or even the Seventh (Dal Corso et al., 2020) Extinction—on the other.

R. Braidotti (✉)
Utrecht University, Utrecht, The Netherlands
e-mail: R.Braidotti@uu.nl

© The Author(s), under exclusive license to Springer Nature Switzerland AG 2023
G. Di Capua and L. Oosterbeek (eds.), *Bridges to Global Ethics*,
SpringerBriefs in Geoethics,
https://doi.org/10.1007/978-3-031-22223-8_7

The combined effects of the posthuman convergence have been exacerbated by the COVID-19 pandemic. Generated by human interference in the environmental and animal world, it proved a powerful catalyst in revealing the under-laying layers and degrees of ecological, social and economic inequality, which the dominant neo-liberal ideology denied. It revealed with incisive cruelty the persistence of patterns of discrimination based on geo-political location, with vaccination, access to advanced technologies, including adequate health and medical care being more readily available in the North than the South of the world. It also brought out persisting inequalities based on class, race, ethnicity, religion, gender and sexuality, able-bodiedness and fluency in dominant language and cultures. These internal fractures and contradictions are emblematic of the posthuman predicament. In this chapter, however, I will not explore these contradictions further, but rather focus on how new materialism in a posthuman frame can help us deal with them. I will stress, especially the importance of renewing our connection to the earth and to the grounded nature of our subjectivity, as a premise to affirmative ethics and inter-species solidarity.

What is "new" about the new materialism today, when compared with earlier historical variations, is a more comprehensive understanding of matter itself— biological, geological, hydrological, meteorological and technological matter. More specifically, contemporary materialism rests on the premise of a "nature-culture" continuum (Haraway, 1985) that is also technologically mediated (Guattari, 2000). This nature-culture-technology continuum forms a transversal connection that bridges the gap between the more traditional dualistic oppositions of nature/culture and technology/matter.

This approach calls for a closer relationship of philosophy and the humanities to scientific culture and especially the Earth and the Life sciences. The materialist turn in contemporary philosophy expands the old distinction between the "two cultures" (Snow [1959] 1998) of the humanities and the sciences to the "three cultures," including contemporary natural and techno-sciences (Kagan, 2009). New materialism constitutes therefore a robust plane of encounter between several disciplinary communities and scholarly traditions (Braidotti, 2019, 2022). One example of these transdisciplinary connections is provided by feminist neo-materialism, with its emphasis on the body, affect and sexuality. Another angle is represented by Spinozist materialism, as re-appraised in French philosophy by Gilles Deleuze (1988, 1990) and Michel Serres (1995). Yet another, ancient tradition is Indigenous epistemologies and cosmologies (Rose, 2004; Viveiros de Castro, 2015), which never rested on a nature-culture divide to begin with. In different but intersecting ways, these alternative philosophical approaches propose a more expanded definition of materialism and matter itself, which includes non-human elements, the dis-qualified or discriminated humans, as well as the technological apparatus. I will not be able to examine all these traditions in this essay, but want to mention them, to argue the main point that what is emerging with posthuman new materialism is a heterogeneous range of process ontologies. They form the basis of a general posthuman ecology (Fuller, 2005, 2008, Horl, 2017) of becoming and affirmative relational ethics (Braidotti, 2006, 2019).

## 7.2   The New-Materialist Turn

Continental philosophy is endowed with a rich tradition of "enchanted materialism" (De Fontenay, [1981] 2001) that is to say a non-deterministic vision of matter, a kind of methodological naturalism that encompasses bodies and sexuality. This corporeal materialism, also known as "the line of immanence," or "affirmative naturalism" (Ansell Pearson, 2014), is specific to French epistemology and philosophy of science. The contemporary new-materialist turn is inspired by this tradition and builds on it. It looks towards the self-organizing vitality of matter, arguing that all living entities are variations on common material forces or elementary particles. Moreover, the vitality of matter has been replicated by the technological apparatus, which is capable of going "live," producing "smart" things and self-correcting artificial intelligence networks.

This enlarged and dynamic—or vital—vision of materialism extends beyond the polarizing oppositions of matter/mind and nature/culture, while it also avoids holistic organicism. The self-organizing capacity of matter is not an undifferentiated mess, but rather a differential system that works by processes of internal differentiation, defined as modulations within a common matter. This view consequently rejects essentialist or flat equivalences across all species, entities, organisms and apparatus. It recognizes instead the importance of differences, but rather in the non-dualistic form of collaborative relational bonds and cross-species interconnections. My approach emphasizes variations in locations, perspectives and degrees of intensity, force and understanding. Philosophical perspectivism is another term that is used to define this philosophical approach that is both grounded and dynamic. The inception of new materialism as embodied, embedded, relational and affective provides the common denominator that connects the human, non-human and dehumanized entities of all species (Braidotti, 2022).

This approach displaces the boundary between *bios*, the portion of life that has traditionally been reserved for Anthropos and *zoe*, the wider scope of animal and non-human life. The dynamic, self-organizing structure of non-human life as *zoe* emerges as generative vitality (Braidotti, 2006, 2011b). It is the transversal force that cuts across and reconnects previously segregated species, categories and domains.

A new-materialist philosophy covers organic non-human life—*zoe*—the geological foundations of living matter—geo—and technological mediation, producing a notion of the subjects as "*zoe*-geo-techno-based" (Braidotti, 2019, 2022). The emphasis on multi-scalar interdependence in a grounded, but also dynamic manner also supports a relational posthuman ethics of affirmation. An ethical position based on these new-materialist premises promotes a cross-species way of thinking that includes the *zoe*-geo-techno-mediated relations to the multiple ecologies composing the posthuman convergence. It foregrounds the material foundations and the relational interconnection of all entities.

In this framework, matter is both situated and fluid, moreover, in a nature-culture technology continuum, it undergoes some paradoxical mutations. For instance, matter is re-materialized by becoming embedded and embodied, for instance in the

ravages visited upon forests, animals and humans alike by environmental depletion, climate change and global pandemics. At the same time, matter is also *de*-
materialized, through advanced technological mediation, which reduces all entities
to quantified data-producing actors, the target of collection and intervention by a
range of information and medical technologies.

A more precise way of expressing this paradoxical situation is to say that the
dichotomy between de/re-materialization is coming undone in so far as, in the present
conjuncture, living entities undergo a material abstraction into another kind of matter.
This is the abstract materiality of digital numbers, storage, codes, data and their technological applications (Fuller, 2005, 2008, 2017). This double pull towards new fluid
forms of re-materialization and de-materialization is constitutive of the posthuman
predicament, in that it results in a double challenge to and displacement of the human.
This occurs firstly through invasive technological intervention and secondly due to
accelerating environmental degradation. It is a paradoxical movement that cannot
be immediately resolved, but must be acknowledged and studied further. My point
is that, accepting to be zoe-geo-techno-matter in a grounded but dynamic manner is
the precondition for a zoe-geo-techno-ethics of mutual interdependence that opens
up to possible alliances and new subject positions. This approach in turn promotes
alternative ethical ways of becoming posthuman.

## 7.3   Critical Spinozism

Selected doses of new materialism emerged within continental philosophy through
the reappraisal of Spinoza by Deleuze and other French philosophers of the 1970s.
They turned to a different political ontology from Marxism after the political
disappointments of the May 1968 insurrection.[1]

Spinoza's central idea is that we, humans and non-humans, are all part of a
common matter or nature, so that there is no mind–body dualism, but rather a
continuum. He postulates a parallelism between mind and matter in a way that is a
great source of inspiration for the current posthuman predicament. The deep anti-
Cartesianism of Spinozist philosophy, that is to say the emphasis on one common
matter and immanence, is suited to the complexity of the contemporary processes of
de/re-naturalization of matter and material entities. It challenges the transcendental
reading of reason and also refutes the dualism that structured Western metaphysics
and its political philosophy.

Matter—and all living entities, humans included—are composed by combinations
of elementary particles and follow the same laws of physics. They are in perpetual
movement, resonate with each other and get self-organized to the extent that they
can be both networked and specified in what theoretical physicist Rovelli defines

---

[1] For the turn to Spinoza in the circles around Louis Althusser, see: Matheron (1969), Deleuze
([1970] 1988, [1968] 1990), Macherey ([1977] 2011), Negri ([1979] 1991), Balibar (1994), Deleuze
and Guattari (1987, 1994), Guattari (1995, 2000), Serres ([1990] 1995).

as: "inextricable complexity." (2014, p. 73). Body and mind are not set in a binary opposition, nor is there a causal connection between them. It is more a matter of productive resonances between them.

New-materialist Spinozist thought argues that we are all part of nature, not in a manner that opposes it to the social but more like a continuum between the two spheres. Leading Spinozist scholar Lloyd (1994, 1996) points out that our relationship to the natural continuum is affected by the historical social context in which we live. Nature is immersed in history and social structures and vice-versa, without dualistic oppositions. Contemporary critical Spinozist materialism avoids dichotomies, by endowing matter with a dynamic principle of relational self-organization, in what is known as a vital process ontology that is socially based and historically conditioned.

This assumption results in building a "vital politics," premised on that continuum (Bennett, 2010; Braidotti, 2002; Sharp & Taylor, 2016). As a consequence, there is no nature-culture divide at the core of our social systems: we are all in this together although we are not all one and the same (Braidotti, 2019, 2022). Materialism is immanent that is to say not an idealized internalization of the outside world through linguistic and cultural representations. There is no such thing as an inert outside-of-the-human—be it soil, body, stone, earthworm, algorithm or code—whose existence depends solely on the activities and perceptions of the human mind. And although matter does get filtered by a linguistic grid and internalized by humans as a psychic representation, it cannot be reduced to this representation.

In other words, matter and thought are different but equal attributes and expressions of the same substance. Spinoza argues for the parallelism of different dimensions: real objects that exist independently of the human mind (realism) and the mental representations of these objects (idealism). What matters is their equivalent and contiguous existence. Re-read with Deleuze, Spinozist materialism entails better knowledge of ethology that is to say the physics of bodies, as well as the philosophical validity of ideas. We are all variations on a common matter although each embodied and embedded entity—including the anthropomorphic ones—has its specific capacities and pleasures.

The immanent, new-materialist worldview demands an adequate—which for me means embodied, embedded, relational and affective—understanding of one's life conditions. This can be achieved through a process of gradual clarification of the ethical forces at play in one's relationship to the said conditions and their affective charges. In this respect, adequate understanding is rational, in the sense of being rigorously argued and not ideological—superstitious, fanatical or delusional. But—I would add—it is also situated, in a zoe-geo-techno-mediated manner. Thinking and living-with others, including human, non-human and technological others, requires a collaborative and relational praxis. It aims at reaching an adequate—rationalist in the relational sense of the term—understanding of the material conditions that structure our subjectivity and connect us to others. Reason is supplemented in this effort by the force of affect and by the imagination.

The impact of critical Spinozist materialism is radical, in that it critiques the liberal notion of individual autonomy, rejecting the transcendental power of consciousness, in favour of a communitarian and environmentally grounded form of transindividuality (Balibar, 1994; Deleuze, 1988). This critique is also helpful in decentring anthropocentrism and human exceptionalism, in a new ethical bond. Affirmative ethics is the establishment of mutually empowering relations based on cooperation and the productive combination of the specific degrees of intensity or *potentia* of each living entity. It aims at increasing each entity's capacity to preserve themselves against adverse forces. Entities and individuals grow thanks to a collaborative ethical and social sense of community. The capacity to resist and fight back against adversity emerges from the same relational capacities that can also potentially cause harm and discomfort to others. Relationality is not intrinsically positive, and that is why we need an ethics of affirmation to guide it towards the social construction of affirmative relations. All we have is others, and relationship to others is constitutive of ourselves: indeed, we are in this together, but we are not one and the same. What binds us together, over and above contractual ties and shared interests, as posited by the limiting framework of legal humanism, is a materially grounded interdependence. This is enhanced by a shared propensity to persevere in our existence and increase our relational capacities, through an ethics of affirmative collaboration.

## 7.4   Elemental Feminist Materialism

At the intersection of feminism with new materialism, some crucial new ideas emerged that inspired a grounded posthuman philosophy. The emphasis falls on the ecological interdependence and cross-species collaboration beyond human exceptionalism. Elemental new-materialist feminists do not think that the proper study of mankind is "Man" or Anthropos. Objects of study include instead: forests, fungi, bacteria, dust and bio-hydro-solar-techno powers, rubbish and high art (Braidotti & Hlavajova, 2018). As I stated above, life is a transversal system of non-human elements, as in animal and vegetable (*zoe*), earthed and planetary matter and relations (*geo*), while it is also technologically mediated (*techno*). The posthuman convergence proposes a terrestrial kind of materialism that is capable of combining a planetary with an earthy or grounded dimension. This change of objects of enquiry also intensifies the studies of technological mediation, with focus not only on language and representation, but also on material algorithms, codes, software, platforms, networks and more (Fuller, 2015). These objects of study are posthuman *zoe*/geo/techno-mediated.

In this heterogeneous framework, matter can be best described as "elemental." The classical elements of earth, air, water and fire have historically inspired quite a philosophical reflection on the materiality of the cosmic and planetary forces, for example the Epicurean and Stoic traditions, which influenced the modern French school of non-deterministic materialism.[2]

An elemental approach was pursued in the French feminist tradition by Luce Irigaray in the first phase of her work: the trilogy on air, about Heidegger (Irigaray, 1999), water, about Nietzsche (Irigaray, 1991) and the essay "the Mechanic of Fluids" (Irigaray, 1985b). She lays the foundations for a materially embodied philosophy of the feminine as not-one, as a figure of complexity who is "otherwise other" than the second sex of "Man" and associated with both natural and cosmic forces (Braidotti, 2002). These materialist aspects of Irigaray's philosophy make her notion of the virtual feminine into a precursor of feminist materialism (Braidotti, 1994; Grosz, 1994; Lorraine, 1999). They resonate especially well with new-materialist Deleuzian feminists (Colebrook, 2000; MacCormack, 2012; Olkowski, 1999). This sexuate materialism has been explored by contemporary feminist scholars who stress the post-anthropocentric and non-binary directions of Irigaray's philosophy (Stone, 2006, 2015; Stark, 2017).

Contemporary materialist feminism has built up the elemental aspects. Stacy Alaimo for instance focusses on water and marine biology, working on oceans, jelly-fish and seas (Alaimo, 2010, 2013, 2016), as does Astrida Neimanis, who coined the term "hydro-feminism" (Neimanis, 2017) and planetary "bodies of water" (Neimanis, 2018, p. 55) to define this new approach. Feminist new-materialist scholars working on water also include Hayward, who studies star-fish (Hayward, 2008, 2012) and Schrader (2012a) who works on algae and slime. The hydrosphere, however, is also technologically mediated, with the distribution of water through multiple urban and rural networks, intersecting with infrastructure and medical bio-technology. Roberts (2008), for instance, follows the course of hormones circulation through and outside the human bodies into the general environment in sewerage systems and leakages.

New materialism in posthuman times brings back the elements to their non-anthropomorphic dimension, through a series of relational assemblages with non-human others. Through the alterations and pressures introduced by the two-pronged impact of technological intervention and climate change, the classical natural elements—fire, earth, air and water—transform themselves into post-natural objects of interaction. Studying these posthuman permutations, Genosko (2018) comments on how the atmosphere, or air, for instance is transformed into greenhouse gases. The geosphere follows suit, with the earth turning into minerals and mines, then dust and particles. The biosphere is all over the place, as fire combusts into ashes, smoke and gas (Protevi, 2013, 2018). Air and the atmosphere, as suggested by Lynn Margulis (Margulis & Sagan, 1995) return in posthuman scholarship as atmospheric pollution. Breathing has become a hot political issue, not only because of climate change, but also due to the Covid-19 epidemic that causes respiratory diseases, but also because of the suffocating effects of racism, xenophobia (Mbembe, 2021) and social hatred of women and LGBTQ+ people.

---

[2] For Deleuze's relationship to Lucretius, see Deleuze (1961, 1966, 1990). For Deleuze's stoicism, see Deleuze (1990, 1994), Ansell Pearson (2014) and Ryan (2020).

New-materialist feminists however focus mostly on the earth as territory, ground, soil and dust that is to say not only as a geological entity but also as a planetary common home (Grosz, 2004). I approached it in terms of materially embodied and embedded relational becoming (Braidotti, 2002). For Povinelli (2017) what is needed now this is a new "geontology," centred on the Earth and the life of the inorganic. Povinelli makes a useful distinction between three instances of gerontology, based on different aspects of geo. Firstly, as a living planetary organism (Gaia); then as the part of the planet defined as inorganic; and finally, as matter, pitched in opposition to the social and bio-political governance. Feminist theorists (Braidotti, 2006; Grosz, 2008; Povinelli, 2016) have turned these geo-centred perspectives into a critique of the anthropocentric bias of bio-centric political theories Special criticism was directed at Foucault's idea of bio-politics and biopower (Foucault, 1977, 1978, 2007), which manage to avoid any reference to the environment and non-human agents. This omission is repaired by the new-materialist turn of Deleuze and Guattari, as I have extensively argued (Braidotti, 2011b, 2013, 2019). Even the established concept of 'biosecurity' has now turned into geo-security and meteo-security, because of the far-reaching effects of climate change.

Yusoff (2018) develops the geo-centred perspective into critical race studies of rocks, minerals, dust and mines. She focusses on the geo-graphical, geological and imperialist dimensions of the climate crisis, which she calls the "geotrauma of the Anthropocene" (Yusoff, 2018, p. 59). Stressing the role that the extraction economies introduced by western colonialism played in the making of the current juncture, Yusoff posits environmental racism as central to the Western world view and the political economies enforced by advanced capitalism.

All these examples indicate that what comes to the fore in contemporary new materialism is a posthuman elemental political subjectivity that is to say the potency and the vitality of living matter itself. The legacy of ecofeminism is foundational here in so far as it stresses the emergence of earthly matter—Gaia—as a political subject within the posthuman predicament. Ecofeminists consequently invented ways of granting political agency to matter and material entities, beyond the sexist and anthropocentric definition of politics and sociality as being inherently and exclusively human, male, heterosexual and Eurocentric (Gaard, 2011; Midgley, 1996; Mies & Shiva, 1993; Plumwood, 1993).

Indigenous and de-colonial populations had been thinking for centuries about the need to live-with and take care of the land, the country, the earth. Simone Bignall draws from Deleuze's reading of Spinoza a potential for a relational ethics of association to inform a postcolonial and pluralist politics of mutuality, in a critical engagement with Australian Aboriginal epistemologies (Bignall & Patton, 2010; Bignall et al., 2016; Bignall & Braidotti, 2019). I will return to this in a later section.

What is distinctive of elemental materialism in the posthuman frame is that attention to the geological elements is not a single issue but becomes a transversal, relational one. This means that we are looking at cross-overs through the politics of strata (Clark, 2008, 2016) and the geopolitics of the physical world (Bonta & Protevi, 2004;

Genosko, 2018; Protevi, 2013). In feminist materialist trans-national politics, planetary emergencies arise from bushfires to the plastic ocean, impacting human and non-human lives.

A focus on elemental materialism illuminates also the extent to which matter can turn into a hermeneutical key for work in the humanities and cultural criticism. Materialist and elemental ecocriticism (Cohen & Duckert, 2018) for instance merges with technology studies, while the environmental roots of digital media become more manifest. These intersections use terms like "cloud" to designate information hubs and the "swarms" of artificial intelligence bytes (Peters, 2015; Tung-Hui, 2015). Arguing for the materially grounded and hence elemental nature of contemporary culture is a basic tenet of posthuman knowledge and of the new humanities, also known as the posthumanities (Braidotti, 2019; Wolfe, 2010). The transversal emphasis on cross-connections is also central to both the mainstream and the feminist posthumanities (Åsberg & Braidotti, 2018; Braidotti, 2019). A new-materialist approach deals with language, literature, art, maths and ideas by focussing on the material foundations and the "natural" elements.

Cultural studies of science and technology and media studies also made quite a crossover into the elements, by paying attention to rocks, dust, waste and minerals (Fuller, 2005, 2008; Gabrys, 2011, 2016, 2020). This shift moves beyond the continuum "nature-cultures" to embrace a materialist cultural geology that focusses on "medianatures" (Parikka, 2015). The method of "geological materialism" reworks the linguistic, semiotic and representational approach to media by focussing on the hardware. Parikka foregrounds the thick materiality of these technologies and the concrete social realities, including labour relations that make media possible in the first place. This transdisciplinary approach centres on the earth's geological formations, the geophysical resources and the geo-political locations of minerals. This allows for a critique of the power relations involved in advanced capitalism, which is in fact a brutal extraction economy based on the deregulation of mines and mining companies that operate with impunity the world over. And because mining takes mostly place on Indigenous and First Nations people's lands with complete neglect of their rights, it revives and exacerbates the colonial legacy (Rose, 2004; Yusoff, 2018). Recent studies look specifically into the issue of post and de-colonial computing (Benjamin, 2019; Chun, 2021; Irani et al., 2012).

In a crossover into computational technologies, Gabrys (2011) presents a "natural history" of electronics in her study of digital waste disposal techniques and practices. She stresses the key role of the minerals that compose the infrastructure and basic components of the deceptively "immaterial" information technologies. By studying how electronic rubbish is broken down, disassembled and stored, from containers to landfills, museums and archives, Gabrys discusses also the labour relations involved in electronic rubbish disposal and exposes the racialized, digital proletariat taking care of this dangerous work. An expert in sensing technologies, Gabrys (2016) also explores citizens' alternative use of technologies to measure climate change and other contested issues. Gabrys' most recent work (2020) analyses the extent to which environmental management—the Internet of trees—is relying on new digital technologies to monitor the progress of "natural" forests. Smart forests result in woods becoming technologized sites of data production, which can also be put to the task of mitigating environmental change and foster well-being.

To conclude this section, feminist new materialism emphasizes elemental bodies as transversal assemblages, which interlinks human bodies with plants, animals, the earth, the weather, technology and other kinds of matter. This is a discourse where *zoe* embraces *geo* and *techno* as interrelated critical dimensions in the contemporary posthumanities.

## 7.5   Racialized Materialism

Vital materialism—and the life philosophies based on it—is not only sexualized, but also racialized and naturalized. The materialist effort, to overcome the binary oppositions nature-culture and the emphasis on cross-species materialist interdependence, can raise the fear of biological determinism and its discriminatory effects. Many critical theorists have stressed the impact of the racialized ontologies that were developed through colonialism and historical European fascism (Smelik & Lykken, 2008; Stoller, 1995, 2002; Jones, 2011).

In order to come to terms with these racialized aspects, we need a critical and non-essentialist philosophy of life that accounts also for the power relations, the exclusions and the genocides. Living in the posthuman convergence demands us to engage not only with the bio-political management of the living, but also the necro-political management of the dying (Foucault, 1977, 1978, 2007; Mbembe, 2003) and to do so in a non-anthropocentric perspective (Haraway, 1997, 2016). Closer links need to be established between critical race theory and science and technology studies, to counter-act the silencing of slavery and colonialism as factors in the history and philosophy of science. Tuana argues that it is urgent to take seriously the racialized roots of materialism as "viscous porosity" (Tuana, 2008) or thick interaction between nature and culture, humans, non-humans and other categories. She calls for a re-materialization of the social and a re-socialization of the natural and stresses the need to recognize the racialization of this interactive philosophy of nature-culture, sociology, biology and geology. And we already saw in the previous section how geo-centred critical theorists approach critical geological studies as the material substratum of colonial economies, slavery and racism. It is practically impossible to disconnect geology, metallurgy and extraction economies from the infrastructure of race, colonialism and slavery.

Working within a non-Western ontology, Danowski and de Castro (2017) foreground the relevance of Indigenous epistemologies and their rejection of the human/animal opposition. Indigenous relationality emphasizes the porous boundaries between categories, the flows of information at all levels between the humans and their multiple non-human environments. Another name for such vital energy is resilience, which is the ability to survive ontologically, as well as individually, against the necro-political aspects of biopower and the violence of sovereign capital.

This inseparable connection between diverse Indigenous world-views and the natural world makes for the method of "Indigenous expressivism," whereby one speaks as country (Bignall et al., 2016). Within the Aboriginal tradition, Rose (2004, p. 153) points out that the notion of land or country is a method as well as the matrix for Australian Aboriginals' world view and their relationship to space and time. Country is a multi-dimensional concept that includes people, rocks, birds, animals and the weather. Embedding human subjects into their environment, Indigenous philosophies posit environmental, social and personal sustainability as a continuum. To destroy the land or the country is to destroy yourself.

Tallbear (2015) argues that even organic matter such as stones, meteorological phenomena like thunder or stars are known within Indigenous ontologies to be sentient and knowing persons. By attributing a soul and agency to all entities, Indigenous cosmologies can be said to "humanize" them, but they do so in a non-anthropocentric and non-hierarchical manner. This is another, far more ancient, form of vital new materialism that intersects with Western philosophical attempts to rethink the unity of matter without deterministic hierarchies. Indigenous approaches moreover foreground the critique of settler colonialism and its violent management of less-than-human and non-human others (Bignal & Rigney, 2019; Viveiros de Castro, 2014).

## 7.6   Affirmative Ethics

I have argued in this chapter that elemental new materialism is foundational for posthuman thought because it offers a dynamic and immanent frame to steer a course between the risks of biological essentialism (total re-materialization) and technological determinism (total de-materialization). The new-materialist process ontology is driven instead by the relational ethics of joy or affirmation. Affirmation is the ethical force that endures and sustains, whereas sad passions bring about impotence and disaggregation. Affirmative ethics is the affect that binds together the heterogeneous components of the complex zoe-geo-techno-subjects.

Affirmative ethics supports a middle ground composed by alliances between human and non-human agents. I have described these alliances as heterogeneous assemblages. To be materialist means to be situated along differential locations: we are grounded, but we flow and we differ. What keeps the human and non-human connected is affect or ontological relationality. This is the capacity to affect others and be affected in mutual interdependence. The interrelation is shaped by an ethics that starts with the recognition of the vital importance of the transversal connections to multiple others. This ethics pursues an affirmative path of relations in order to increase the mutual and respective ability of materializing the unrealized or virtual potential of what "we" are capable of becoming.

New materialism entails several consequences for posthuman affirmative ethics. Methodologically, it produces a post-naturalist but also post-constructivist form of embodied and embedded empiricism. There are no humanist subjects of transcendental reason in posthuman thought: no self-referential entities locked up in their own exceptionalism. The subjects at work in these assemblages are rather immanent, transversal, heterogeneous and collaborative. They are rooted in embodied and embedded locations but flowing in a multi-layered intersectional manner. The subjects of posthuman affirmative ethics do not correspond to the unitary vision of "Man/Anthropos" that regulates access to the dominant category of the human. They are post-anthropocentric in the sense of elemental materialism and interconnected with non-human agents and forces.

This is not to deny the specificity of humans, but rather to challenge the claims to human supremacy. For humans, as a specific embodied and embedded, genetic and neural entity, any self-understanding is necessarily anthropomorphic. That is to say that it is characterized by the qualities, faculties and potentials of the "embrained bodies and embodied brains" that we happen to be. Given the corporeal materialism that is specific to our species, human beings cannot help but be anthropomorphic and although we are diverse and differentiated, we do share some basic parameters: morphological, cellular, neural and hormonal. Posthuman affirmative ethics embraces anthropomorphic specificity, in order to take distance from the arrogance of anthropocentrism and human exceptionalism. The acknowledgement of the immanence and grounded contingency of just a life inoculates against the intoxicating metaphysics of *Life* itself. Paradoxically, it is by embracing resiliently the specificity of the anthropomorphic frame and the limits and possibilities it entails that humans can connect to non-human others. Being grounded, we can flow across to others more easily. In this ethical relational effort, the shared imaginary and experiences of the less-than-human and dehumanized others, who were not allowed to be fully human to begin with, are inspirational. Accepting to be zoe-geo-techno-matter is a way of opening up to possible alliances, which in turn promotes alternative ways of becoming posthuman.

Affirmative ethics acknowledges the shared desire of all entities to persevere in their collaborative interdependence and to increase it for the common good. In this sense, affirmation is not a psychological disposition, but an energetic enhancement that marks a collective increase in our relational ability to take in and take on more of the world. Defining affirmative ethics as the establishment of mutually empowering relationships based on cooperation and the combination of the specific powers of each entity aims at increasing each entity's individual capacity to preserve themselves against adverse forces. Entities and individuals grow thanks to a collaborative connection. The capacity to resist and fight back emerges from the same relational capacities that can also potentially cause harm and discomfort: all we have is others and relationship to others is constitutive of ourselves. What binds us together over and above contractual interests and the limiting constraints of humanism is a common propensity to persevere in our existence and increase our relational capacities. An ethics of affirmative collaboration is our common factor.

Affirmation is a praxis that needs to be constructed by avoiding binary polarizations and suspending judgement. A respect for complexity needs to be developed, stressing the inevitable fact that contradictions are part of any worthy political and intellectual project. That the truth is not clear cut and is not a matter of "either/or" but rather of "and…and." Affirmative ethics is about relational interconnections, pacifism, non-violence and generosity.

It is an affirmative ethical praxis that aims to cultivate and compose a new collective subject. This subject is an assemblage—"we"—that is a mix of humans and non-humans, zoe/geo/techno-bound, computational networks and earthlings, linked in a vital interconnection that is smart and self-organizing, but not chaotic. Let us call it, for lack of a better word, "a life."

Death is an essential part of it. So, many lives today are the object of threatening death-forces and new ways of dying, on a planetary scale, as a result of wars, migrations, pandemics, viruses and other devastating events. And many of these endangered lives are not human; many are unreported or even unregistered.

This is the insight of posthuman thought as a secular, new-materialist eco-philosophy of becoming. Life is a generative force beneath, below, and beyond what we humans have made of it. It is an inexhaustible zoe-geo-techno-mediated and generative force that constitutes a life, in its complexity and interdependence. The multiplicity of singular but interconnected lives has the potential to generate affirmative relations with all others, with all that all lives and endures, also the non-human.

# References

Alaimo, S. (2010). *Bodily natures*. Indiana University Press.
Alaimo, S. (2016). *Exposed*. University of Minnesota Press.
Alaimo, S. (2013). Jellyfish science, jellyfish aesthetics: Posthuman reconfigurations of the sensible. In J. MacLeod, C. Chen, & A. Neimanis (Eds.), *Thinking with water*, McGill-Queens University Press
Ansell-Pearson, K. (2014). Affirmative naturalism: Deleuze and Epicureanism. *Cosmos and History, 10*(2), 121–137.
Asberg, C., & Braidotti, R. (Eds.). (2018). *A feminist companion to the Posthumanities*. Springer International Publishing.
Balibar, E. (1994). *Spinoza and politics*. Verso.
Benjamin, R. (2019). *Race after technology*. Polity Press.
Bennett, J. (2010). *Vibrant matter*. Duke
Bignall, S., & Patton, P. (2010). *Deleuze and the postcolonial*. Edinburgh University Press.
Bignall, S., Hemming, S., & Rigney, D. (2016). Three ecosophies for the Anthropocene: Environmental governance, continental posthumanism and indigenous expressivism. *Deleuze Studies, 10*(4), 455–478.
Bignall, S., & Rigney, D. (2019). Indigeneity, posthumanism and nomad thought: Transforming colonial ecologies. In R. Braidotti, & S. Bignall (Eds.), *Posthuman ecologies*. Rowman & Littlefield.
Bonta, M., & Protevi, J. (2004). *Deleuze and geophilosophy. A guide and glossary*. Edinburgh University Press.
Braidotti, R. (2002). *Metamorphoses: Towards a materialist theory of becoming*. Polity Press.
Braidotti, R. (2006). *Transpositions: On nomadic ethics*. Polity Press.
Braidotti, R. ([1994] 2011a). *Nomadic subjects. Embodiment and sexual difference in contemporary feminist theory*. Columbia University Press

Braidotti, R. (2011b). *Nomadic theory*. Columbia University Press.

Braidotti, R. (2013). *The Posthuman*. Polity Press.

Braidotti, R. (2019). *Posthuman knowledge*. Polity Press.

Braidotti, R. (2022). *Posthuman feminism*. Polity Press.

Braidotti, R., & Hlavajova, M. (Eds.). (2018). *Posthuman glossary*. Bloomsbury Academic.

Chun, W. (2021). *Discriminating data*. MIT Press.

Clark, N. (2008). Aboriginal cosmopolitanism. *International Journal of Urban and Regional Research, 32*(3), 737–744. https://doi.org/10.1111/j.1468-2427.2008.00811.x

Clark, N. (2016). Politics of strata. *Theory, Culture & Society, 34*(2–3), 211–231. https://doi.org/10.1177/0263276416667538

Cohen, J. J., & Duckert, L. (2018). *Elemental ecocriticism*. University of Minnesota Press.

Colebrook, C. (2000). Is sexual difference a problem? In I. Buchanan, & C. Colebrook (Eds.), *Deleuze and feminist theory*. Edinburgh University Press.

Dal Corso, J., Bernardi, M., Sun, Y., Song, H., & Seyfullah L. J. et al. (2020). Extinction and dawn of the modern world in the Carnian (Late Triassic). *Science Advances, 6*(38), eaba0099. https://doi.org/10.1126/sciadv.aba0099

Danowski, D., & Viveiros de Castro, E. (2017). *The ends of the world*. Polity Press.

Deleuze, G. (1961). Lucrèce et le naturalisme. *Les Etudes Philosophiques, 16*(1), 19–29.

Deleuze, G. (1966). Renverser le platonisme (Les simulacres). *Revue De Métaphysique Et De Morale, 71*(4), 426–438.

Deleuze, G. (1994). *Difference and repetition*. Athlone Press.

Deleuze, G., & Guattari, F. (1987). *A thousand plateaus: Capitalism and schizophrenia*. University of Minnesota Press.

Deleuze, G., & Guattari, F. (1994). *What is philosophy?* Columbia University Press.

Deleuze, G. ([1968] 1990). *Expressionism in philosophy: Spinoza*. Zone Books.

Deleuze, G. ([1970] 1988). *Spinoza: Practical philosophy*. City Lights Books.

Horl, E. (Ed.). (2017). *General ecology*. Bloomsbury Academic.

De Fontenay, E. ([1981] 2001). Diderot ou le Matérialisme Enchanté. Paris: Grasset.

Foucault, M. (1977). *Discipline and punish*. Pantheon Books.

Foucault, M. (2007). *The Birth of biopolitics: Lectures at the Collège De France, 1978–79*. Palgrave.

Foucault, M. (1978). *The will to knowledge. The history of sexuality: 1*. Penguin Books.

Fuller, M. (2005). *Media ecologies: Materialist energies in art and technoculture*. MIT Press.

Fuller, M. (2008). *Software studies: A lexicon*. MIT Press.

Fuller, M. (2017). *How to be A Geek. Essays on the culture of software*. Polity Press.

Gaard, G. (2011). Ecofeminism revisited: Rejecting essentialism and re-placing species in a material feminist environmentalism. *Feminist Formations, 23*(2), 26–53. https://doi.org/10.1353/ff.2011.0017

Gabrys, J. (2011). *Digital rubbish: A natural history of electronics*. University of Michigan Press.

Gabrys, J. (2016). *Program earth: Environmental sensing technology and the making of a computational planet*. University of Minnesota Press.

Gabrys, J. (2020). Smart forests and data practices: From the Internet of Trees to planetary governance. *Big Data & Society*, January–June: 1–10. https://doi.org/10.1177/2053951720904871

Genosko, G. (2018). Four elements. In R. Braidotti & M. Hlavajova (Eds.), *Posthuman glossary* (pp. 167–169). Bloomsbury Academic.

Grosz, E. (1994). *Volatile bodies. Towards a corporeal feminism*. Indiana University Press.

Grosz, E. (2004). *The nick of time*. Duke University Press.

Grosz, E. (2008). *Chaos, territory, art. Deleuze and the framing of the earth*. Columbia University Press.

Guattari, F. (1995). *Chaosmosis. An ethico-aesthetic paradigm*. Power Publications.

Guattari, F. (2000). *The three ecologies*. The Athlone Press.

Haraway, D. (2016). *Staying with the trouble: Making kin in the chthulucene*. Duke University Press.

Haraway, D. (1985). A Manifesto for Cyborgs: Science, technology, and socialist feminism in the 1980s. *Socialist Review, 80,* 65–108. http://www.f.waseda.jp/sidoli/Haraway_Cyborg_Man ifesto.pdf. Accessed September 21, 2022.

Haraway, D. (1997). Modest_Witness@Second_Millennium. In *FemaleMan©_Meets_ Oncomouse.* Routledge.

Hayward, E. (2012). Sensational jellyfish: Aquarium affects and the matter of immersion. *Differences, 23*(3), 161–196. https://doi.org/10.1215/10407391-1892925

Hayward, E. (2008). More lessons from a starfish: Prefixial flesh and transspeciated selves. *Women's Studies Quarterly, 36*(3/4), 64–85. https://static1.squarespace.com/static/5f19ed038 2525a1ca134124a/t/5f7d0d9fecabfd5e06def5b6/1602031008659/More+Lessons+From+a+Sta rfish+-+Eva+Hayward+%281%29.pdf. Accessed September 21, 2022.

Sharp, H., & Taylor, C. (Eds.). (2016). *Feminist philosophies of life.* McGill-Queen's University Press.

Irani, L., Vertesi, J., Dourish, P., & Kavita, P. (2012). Postcolonial computing. A tactical survey. *Science, Technology, & Human Values, 37*(1), 3–29. https://doi.org/10.1177/0162243910389594

Irigaray, L. ([1980] 1991). *Marine lover.* Columbia University Press.

Irigaray, L. ([1974] 1985). *Speculum of the other woman.* Cornell University Press.

Irigaray, L. ([1983] 1999). *The forgetting of air in Martin Heidegger.* University of Texas Press.

Jenkins, B. (2017). *Eros and economy: Jung, deleuze, sexual difference.* Routledge.

Jones, R. (2011). *Irigaray: Towards a sexuate philosophy.* Polity Press.

Kagan, J. (2009). *The three cultures: Natural sciences, social sciences and the humanities in the 21st Century.* Cambridge University Press.

Kolbert, E. (2014). *The sixth extinction.* Henry Holt Company.

Lloyd, G. (1994). *Part of nature: Self-knowledge in Spinoza's ethic.* Cornell University Press.

Lloyd, G. (1996). *Spinoza and the ethics.* Routledge.

Lorraine, T. (1999). *Irigaray & Deleuze: Experiments in visceral philosophy.* Cornell University Press.

MacCormack, P. (2012). *Posthuman ethics.* Ashgate.

Macherey, P. ([1977] 2011). *Hegel or Spinoza?* University of Minnesota Press.

Margulis, L., & Sagan, D. (1995). *What is life?* University of California Press.

Matheron, A. (1969). *Individu et communauté chez Spinoza.* Les Editions de Minuit.

Mbembe, A. (2003). Necropolitics. *Public Culture, 15*(1), 11–40. https://doi.org/10.1215/089 92363-15-1-11

Mbembe, A. (2021). The universal right to breathe. *Critical Inquiry, 47*(S2), 58–62. https://doi.org/ 10.1086/711437

Midgley, M. (1996). *Utopias, dolphins and computers. Problems of philosophical plumbing.* Routledge.

Mies, M., & Shiva, V. (1993). *Ecofeminism.* Zed Books.

Negri, A. ([1979] 1991). *The savage anomaly: The power of Spinoza's Metaphysics and Politics.* University of Minnesota Press.

Neimanis, A. (2017). *Bodies of water: Posthuman feminist phenomenology.* Bloomsbury.

Neimanis, A. (2018). Posthuman phenomenologies for planetary bodies of water. In C. Asberg & R. Braidotti (Eds.), *A feminist companion to the Posthumanities* (pp. 55–66). Springer.

Olkowski, D. (1999). *Gilles Deleuze and the ruin of representation.* University of California Press.

Parikka, J. (2015). *A geology of media.* University of Minnesota Press.

Peters, J.D. (2015). *The marvelous clouds: Towards a philosophy of elemental media.* The University of Chicago.

Plumwood, V. (1993). *Feminism and the mastery of nature.* Routledge.

Povinelli, E. (2016). *Geontologies: A requiem to late liberalism.* Duke University Press.

Povinelli, E. (2017). The three figures of Geontology. In R. Grusin (Ed.), *Anthropocene feminism.* University of Minnesota Press.

Protevi, J. (2013). *Life, war, earth.* University of Minnesota Press.

Protevi, J. (2018). Geo-hydro-solar-bio-techno-politics. In R. Braidotti, & M. Hlavajova (Eds.), *Posthuman glossary*. Bloomsbury Academic.

Roberts, C. (2008). Fluid ecologies. Changing hormonal systems of embodied difference. In A. Smelik, & N. Lykke (Eds.), *Bits of life: Feminism at the intersections of media, bioscience and technology*. University of Washington Press.

Rose, D. B. (2004). *Reports from a wild country*. University of New South Wales Press.

Ryan, J. J. (2020). *Deleuze, A Stoic*. Edinburgh University Press.

Schrader, A. (2012). The time of slime: Anthropocentrism in harmful algal research. *Environmental Philosophy, 9*(1), 71–94. https://doi.org/10.5840/envirophil2012915

Schwab, K. (2015). The fourth industrial revolution. Foreign Affairs, December 12. https://www.foreignaffairs.com/world/fourth-industrial-revolution. Accessed September 21, 2022.

Serres, M. ([1990] 1995). *The natural contract*. University of Michigan Press.

Smelik, A., & Lykke, N. (Eds.). (2008). *Bits of life: Feminism at the intersections of media, bioscience and technology*. University of Washington Press.

Snow, C. P. ([1959] 1998). The two cultures. Cambridge University Press.

Stark, H. (2017). *Feminist theory after deleuze*. Edinburgh University Press.

Stoller, A. (1995). *Race and the education of desire*. Duke University Press.

Stoller, A. (2002). *Carnal knowledge and imperial power*. University of California Press.

Stone, A. (2006). *Luce Irigaray and the philosophy of sexual difference*. Cambridge University Press.

Stone, A. (2015). Irigaray's ecological phenomenology: Towards an elemental materialism. *Journal of the British Society for Phenomenology, 46*(2), 117–131. https://doi.org/10.1080/00071773.2014.960747

Tallbear, K. (2015). An indigenous reflection on working beyond the human/not human. *GLQ, 21*(2–3), 230–235.

Tuana, N. (2008). Viscous porosity: Witnessing Katrina. In A. Stacy, & S. Hekman (Eds.), *Material feminisms*. Indiana University Press.

Tung-Hui, H. (2015). *A prehistory of the cloud*. MIT Press.

Viveiros de Castro, E. (2014). *Cannibal metaphysics*. Univocal Publishing.

Viveiros de Castro, E. (2015). *The relative native: Essays on indigenous conceptual worlds*. HAU Press.

Wolfe, C. (2010). *What is Posthumanism?* University of Minnesota Press.

Yusoff, K. (2018). *A billion black anthropocenes or none*. Verso Books.

# Chapter 8
# Epilogue

**Torbjörn Lodén**

**Abstract** Scientific and technological development, economic growth and global-ization have brought about enormous improvements in the quality of life for people all over the world. But these improvements have been costly in terms of global warming, environmental destruction and increasing international tensions. As a reaction we may now observe in many countries an inward turn away from the world, which sees influence from "foreign" countries and cultures as the root of the problems. But today's global problems require global solutions, and to accomplish such solutions humanity needs a global ethic that sees the human being as part of a larger context together with animals, plants, rivers and mountains. Seeking to formulate such a global ethic, geoethics may make decisive contributions.

**Keywords** Environmental destruction · Globalization · Universal values · Modernization · Global ethics

## 8.1 Introduction

Some of today's most serious problems such as global warming and environmental destruction affect all humankind as well as animals and plants. As never before in human history, the global ecosystem is now threatened. These problems are related to economic growth and to globalization.

T. Lodén (✉)
Professor Emeritus at Stockholm University, Stockholm, Sweden
e-mail: luoduobi@gmail.com

Royal Swedish Academy of Letters, History and Antiquities, Stockholm, Sweden

Stockholm China Center at the Institute for Security & Development Policy (ISDP), Stockholm, Sweden

Economic development and modernization have resulted in dramatically improved standards of living for people all over the world. But the generated wealth has been unequally distributed both within countries and among different countries. Thus, inequality has resulted both in social problems in many countries and in increasing international tension. Urbanization and migration have been accompanied by the destruction of traditional family structures, rising crime rates, insecurity, sense of lost identity. No matter how much we value the blessings that modernization has brought about in terms of personal freedom, material welfare, cultural richness, educational standards, yet we cannot deny that modernization has also had and continues to have a destructive side.

Globalization has brought about remarkable results. As a China scholar, I have observed how the opening up of China has contributed to extraordinary economic growth and to greatly improved quality of life for hundreds of millions of people in China. This improvement has by no means been confined to the material standard of living but also manifested itself in cultural richness and diversity and greatly improved educational standards. In spite of the continued, and in recent years even increasingly authoritarian rule in China, hundreds of millions of people have much greater freedom today than before the opening up to decide about their own lives—where to live, what kind of profession to choose, whom to marry. As the work of Hans Rosling et al. (2018) have shown, modernization and globalization have indeed resulted in greatly improved quality of life for the great majority of people on our planet.

But globalization has also further aggravated the negative effects of modernization just mentioned. For example, it cannot be denied that it has contributed to the rapidly increasing unequal distribution of wealth (Piketty, 2014). Moreover, globalization has facilitated the spread of viruses around the world and thereby the risk of outbreaks of pandemics such as COVID-19. In terms of war and peace, many of us were convinced that globalization would serve peace. It would promote cross-cultural understanding and thereby mutual understanding, and it would create bonds that would make states interdependent and therefore less prone to fight one another. Sadly enough, the world has not become more peaceful in recent years, rather the contrary. When writing this, the war is still raging in Ukraine, and we can see how it not only brings devastation and suffering to the country and people of Ukraine, but also affects the global food situation, since Ukraine is a major supplier of grain in the world (Caprile, 2022).

It would be wrong to dismiss the idea that globalization may serve peace, but we must admit that there is no iron logic that this is not always and necessarily so. While globalization makes countries more interdependent, it also results in more opportunities for controversies and conflicts. In terms of war and peace it is a double-edged sword.

As for global warming and the ecological crisis, we can see that in spite of considerable efforts to reach agreements on measures that would halt the rapidly moving downward spiral, manifested for example in the Paris Agreement of 2015 on climate change, these efforts have in fact so far failed to reach the set targets and the globe is fast moving towards a point of no return. In some areas this point may have already been passed.[1]

## 8.2 The Inward Turn

In many countries all over the world, one response to today's problems is to see them as caused by hostile external forces. In my own country Sweden, it has become common for politicians to speak about "Swedish values" rather than "human" or "universal values." All over Europe, it is fashionable to express one's support for the values of one's own country or culture as opposed to values and ways of life that are "foreign." In the USA Donald Trump and his Make America Great Again (MAGA) ideology inspires millions of people. In China, emphasizing "the specific characteristic of Chinese culture" has become an integral part of the dominant discourse. The examples of this kind of inward turn could easily be multiplied. It has become a global trend, and nationalism has come to overshadow internationalism and cosmopolitanism as a major orientation.

But the problems are global and they need global solutions. Turning inwards may bring short-time security, but this cannot be the road ahead towards a peaceful world, where people and cultures live in harmony, while pursuing sustainable ways of life. Global solutions are a matter of the survival of human civilization.

## 8.3 Global Solutions Require Global Ethics

To achieve global solutions, consensus across national and cultural borders is essential. This in turn requires much more of cross-cultural understanding based on a truly global outlook than there is in today's world, and one key element of such a global outlook must be a set of universally accepted ethical norms, a global ethic. The Universal Declaration of Human Rights adopted by the United Nations[2] in 1948 represents an important step in the direction of codifying a set of such norms. One important aspect of this Declaration is that it was rooted in different cultural traditions, not only in Western traditions. The committee that authored the Declaration had an impressively cosmopolitan composition. It was headed by the American Eleanor Roosevelt and also included John P. Humphrey of Canada, René Cassin of France,

---

[1] Concerning the Paris Agreement, see "What is the Paris Agreement?" published by the United Nations: https://unfccc.int/process-and-meetings/the-paris-agreement/the-paris-agreement (accessed 21 September 2022). For this year's IPCC report, see https://journeytozerostories.neste.com/sustainability/ipcc-report-analysis-top-five-measures-halve-emissions-2030?gclid=CjwKCAjwyaWZBhBGEiwACslQo1ug_-YuoeIn-S6SCRQNusCftLDB5i1OE0XmBxj5KaCGaumPNxRd5xoColYQAvD_BwE (accessed 21 September 2022).

[2] https://www.un.org/en/about-us/universal-declaration-of-human-rights (accessed 21 September 2022).

Charles Malik of Lebanon, P. C. Chang of China and Hansa Jivrai Mehta of India. The Chinese scholar and diplomat P. C. Chang insisted that the Declaration should be anchored not only in the Western tradition but also in Chinese tradition and especially in the legacy of Confucian thought that he associated primarily with the philosopher Mencius (Roth, 2018).

The notion of fundamental, universally valid ethical norms is ancient and must, as far as I can understand, be considered an essential aspect of ethical thought that transcends both geographical and temporal boundaries. But the meaning and function of this notion have varied immensely both within one and the same tradition and between different traditions. In the post-World War II era, the idea of universal values, especially in so far as human rights are concerned, has often been described as a tool used by the major Western Powers to oppress other countries and maintain its cultural and ideological hegemony in the world. One cannot say that there has been no basis whatsoever for this accusation, but this need not and should not be so. The notion of universal values, like so many other philosophical notions, can be used in different ways and for different purposes. This is one reason why it is important that the right to interpret what these values mean must be open to free and unfettered discussion. It is essential to define our notion of universal values in such a way that they include rather than preclude respect for difference. In this regard there is a wonderful passage in the Zhuangzi, the famous Daoist book, that can inspire us:

> Once a seabird alighted in the suburbs of the Lu capital. The Marquis of Lu escorted it to the ancestral temple, where he entertained it, performing the Nine Shao music for it to listen to and presenting it with the meat of the Tailao sacrifice to feast on. But the bird only looked dazed and forlorn, refusing to eat a single slice of meat or drink a cup of wine, and in three days it was dead. This is to try to nourish a bird with what would nourish you instead of what would nourish a bird. If you want to nourish a bird with what nourishes a bird, then you should let it roost in the deep forest, play among the banks and islands, float on the rivers and lakes, eat mudfish and minnows, follow the rest of the flock in flight and rest, and live any way it chooses. [...] Fish live in water and thrive, but if men tried to live in water they would die. Creatures differ because they have different likes and dislikes. Therefore, the former sages never required the same ability from all creatures or made them all do the same thing.[3]

The Universal Declaration is a very important document, but it does not play as an important role as many of us would like to see. The inward turn and growing narrow nationalism breeds scepticism with regard to transnational agreements and global norms. Also examples of mismanagement and even corruption in the work of the United Nations and other international organizations provide fuel not to accept such norms. This is very unfortunate. The operations of the United Nations leave much to be desired, but the organization also does a lot of good. The IPPC panel, Agenda 2030 are but two of many examples of its extremely important work.[4] Some of the

[3] From: *The complete works of Zhuang Zi*. trans. Burton Watson, New York: Columbia University Press, 2013: https://terebess.hu/english/tao/Zhuangzi-Burton-Watson.pdf (accessed 21 September 2022).

[4] Concerning Agenda 2030, see "Transforming our World: the 2030 Agenda for Sustainable Development" issued by the UN Department of Economic and Social Affairs, https://sdgs.un.org/2030agenda (accessed 21 September 2022).

UN's problems are probably rooted in its basic structure, which after being in place for soon 80 years is in need of revision. Mismanagement and corruption are serious problems that must be dealt with but certainly no reasons to turn away from the organization.

## 8.4 The Need to See Humans as an Integral Part of a Larger Whole

One prominent feature of modernity as it has evolved since the scientific revolution and the emergence of industrial society is the beginning liberation from structures and ideas that have in various ways restricted us and also, although to a lesser degree, continues to restrict us. As Immanuel Kant put it in calling for enlightenment, rulers have often seen people as mere elements in the kingdom of nature, but to become enlightened, realize their humanity and become the masters of their own destinies people must enter "the kingdom of freedom and ends."[5] In this vein traditional ideas that deified mountains and rivers and called for respect for nature were rejected, not without reason, as superstitious. But maybe this was to throw out the baby with the bathwater, since it meant a departure from a holistic perspective of man and nature.

The changes that modernization has brought about are in many ways a blessing, and the objective of realizing our rationality and seeking to fully enter the kingdom of freedom and ends is an ideal that will always be worth to pursue. But we can also see that the notion that we humans are the masters over nature has legitimized the use of our natural resources to attain material wealth without giving proper consideration to the consequences for the environment. The purpose has not been to cause environmental destruction, but in many cases, this has still been the effect.

When we think about global ethics, we include the world beyond humans: animals and plants, seas and mountains. The tremendous growth of scientific knowledge that has been achieved in the modern era must be put to use in a holistic perspective, and global ethics must be formulated within such a broad perspective. As the climate crisis becomes more and more acute, an increasing number of people also realize that this is the case. One significant example of this Pope Francis' encyclical letter "Laudato Si' on Care for our Common Home" of 2015.[6] But even long before the climate crisis became acute, there were influential thinkers in the modern era who argued that ethics must be defined in a broad perspective where humans are seen as forming a unity with nature. For me as a Scandinavian it is close at hand to think of the Norwegian

---

[5] See Immanuel Kant, "Was ist Aufklärung?" first published as "Was ist Aufklärung?" Berlin-ische Monatsschrift, 1784, s. 481–494. Available on the net: https://www.rosalux.de/filead min/rls_uploads/pdfs/159_kant.pdf (accessed 21 September 2022). English translation "What is Enlightenment?" https://resources.saylor.org/wwwresources/archived/site/wp-content/uploads/2011/02/What-is-Enlightenment.pdf (accessed 21 September 2022).

[6] See "Encyclical Letter Laudato Si' of the Holy Father Francis on care for our common good". Digital edition: https://www.vatican.va/content/francesco/en/encyclicals/documents/papa-francesco_20150524_enciclica-laudato-si.html (accessed 21 September 2022).

philosopher Arne Naess (1912–2009), who first gained fame as a philosopher of the analytic tradition and only as an already celebrated analytic philosopher launched his "ecophilosophy" or "ecosophy." Naess developed his ecophilosophical ideas on the basis of his familiarity with nature (he was himself an accomplished mountaineer), his awareness of the growing environmental problems as well as his reading of the history of philosophy. Spinoza was one of his favourite philosophers, and in his own History of Philosophy he says that Spinoza held that "one should […] see man as a fraction of the whole – partaking in, and dependent on, the laws that apply everywhere and always" (Naess, 1962). Not only a thinker but also an ecological activist, Naess was convinced that one important factor behind the ongoing environmental destruction was the failure of people in the modern era to understand the relationship between humans and nature. He treasured the enlightenment values and modernity, but he deplored the fact that the veneration of nature that had been a prominent feature of much of premodern thought seemed to have disappeared with modernization.

In recent years geoethics has emerged as a new current of ecological and ethical thought (Peppoloni & Di Capua, 2020, 2022). In this volume, to which I am honoured to have been asked to add a short epilogue, the editors have assembled six chapters that demonstrate the vitality of this new current. These chapters give fascinating insights into what geoethics is and which could be its further development. It would be far beyond my competence to try to discuss the meaning and significance of geoethics, but I still wish to point to one feature that strikes me when reading through the six chapters: geoethics is not made up of a closed set of doctrines. It is very much a vital and ongoing, open-ended discussion within a framework that seeks to bring together knowledge from different branches of learning so as to offer a perspective of human beings as "fractions of the whole," to use the words of Arne Naess.

## 8.5   Concluding Words

Modernity based on scientific and technological development and enlightenment thought have brought about enormous progress and improvements for billions of people, and in many ways this progress continues.[7] Yet, humankind now also faces problems that are so serious that they pose a serious threat to the future of human civilization. To these problems belong the climate and environmental crises, deteriorating international relations and military conflicts, oppression, exploitation. In order to preserve civilization and to go further on the path towards a life of material security and cultural richness for all and the respect of everybody's inalienable dignity and rights, it is necessary to deal with these problems but also to reconsider our philosophical and ethical underpinnings. In this era of globalization, to seek consensus

---

[7] See, e.g., this year's report from the Bill and Melinda Gates Foundation: https://www.gatesfoundation.org/goalkeepers/report/2022-report/ (accessed 21 September 2022). See also an interview with Bill Gates, The World is Really Getting Better, The Atlantic, 13 September, 2022: https://www.theatlantic.com/newsletters/archive/2022/09/bill-melinda-gates-foundation-goalkeepers-report-poverty/671415 (accessed 21 September 2022).

about a holistic global ethic rooted in different cultural traditions should, as I see it, be a major priority. As a student of Chinese culture, I believe that Chinese cultural traditions may offer us important resources for such a global ethic. In a recent article I have, as examples of such sources, listed the following features that we can find in the Confucian tradition: humanism, holism and universalism, the search for unity and harmony, the emphasis on responsibility and duties, which as part of a global ethic should be combined with the value of freedom, and finally the emphasis on study.[8] We may note that these features are all very close to the gist of geoethics (Peppoloni & Di Capua, 2020).

There is no doubt that other cultural traditions, such as Hinduism and Buddhism, Islam Judaism, African and American traditions as well as the cultures of indigenous peoples in different parts of the world are also rich depositories of sources for a holistic global ethic. Contemporary ethical and philosophical thought also offer important sources, and the UN Universal Declaration of human rights remains a text of central significance. As this collection of papers shows, geoethics is now emerging as a most promising current in the ongoing quest for a global ethic for our time.

# References

Caprile, A. (2022). *Report for the European Parliament "Russia's War on Ukraine: Impact on Food Security and EU Response"*. https://www.europarl.europa.eu/RegData/etudes/ATAG/2022/729367/EPRS_ATA(2022)729367_EN.pdf. Accessed September 21, 2022.

Naess, A. (1962). Filosofins histoire, Vol. 2, Oslo: Universitetsforlaget, p. 115. Translation from the Norwegian original by the author of this chapter.

Peppoloni, S., & Di Capua, G. (2020). Geoethics as global ethics to face grand challenges for humanity. In G. Di Capua, P. T. Bobrowsky, S. W. Kieffer, & C. Palinkas (Eds.), *Geoethics: Status and future perspectives* (Vol. 508, pp. 13–29). Geological Society of London, Special Publications. https://doi.org/10.1144/SP508-2020-146

Peppoloni, S., & Di Capua, G. (2022). Geoethics: Manifesto for an ethics of responsibility towards the Earth. *Springer, Cham.* https://doi.org/10.1007/978-3-030-98044-3

Piketty, T. (2014). *Capital in the twenty-first century.* Belknap Press of Harvard University Press.

Rosling, H., Rosling, O., & Rosling R. A. (2018). Factfulness: Ten reasons we're wrong about the world. London: Sceptre, http://ak.sbmu.ac.ir/uploads/Fact_Fullness_Hans_Rosling.pdf. Accessed September 21, 2022.

Roth, H. I. (2018). *P.C. Chang and the universal declaration of human rights.* University of Pennsylvania Press.

---

[8] *Confucianism and the Global Challenges of the Twenty-first Century*, to be included in Chun-chieh Huang and John Tucker, eds., Confucianism for the Twenty-First Century, Göttingen: Vandenhoeck & Ruprcht, expected to come out in 2023.

Printed in the United States
by Baker & Taylor Publisher Services